Certified Flight Instructor Oral Exam Guide

Third Edition

The comprehensive guide to prepare you for the FAA Oral Exam

by Michael D. Hayes

Aviation Supplies & Academics, Inc.
Newcastle, Washington

Certified Flight Instructor Oral Exam Guide
Third Edition

Aviation Supplies & Academics, Inc.
7005 132nd Place SE
Newcastle, Washington 98059-3153

ISBN 1-56027-378-X
ASA-OEG-CFI3

Printed in the United States of America

04 03 02 01 00   9 8 7 6 5 4 3 2

Library of Congress Cataloging-in-Publication Data:

Hayes, Michael D.
   Certified flight instructor oral exam guide : the comprehensive guide to prepare you
for the FAA oral exam / by Michael D. Hayes.
        p.   cm.
   Includes bibliographical references.
   ISBN 1-56027-194-9
   1. Flight training—Examinations, questions, etc.—Study guides.
2. Airplanes—Piloting—Examinations—Study guides. 3. United States, Federal
Aviation Administration—Examinations—Study guides.   I. Title
TL712.H39 1994   94-42322
629.132'52'0715—dc20   CIP

This guide is dedicated to the many talented students, pilots, and flight instructors I have had the opportunity to work with over the years. Also special thanks to Mark Hayes, David Sickler, the staff at Howell Aircraft Service, and the many others who supplied the patience, encouragement, and understanding necessary to complete the project.

—M.D.H.

Contents

Continued

9 Ground Reference Maneuvers

10 Approaches and Landings

Introduction

The *Certified Flight Instructor Oral Exam Guide* is a comprehensive guide designed for commercial pilots who are involved in training for the initial Flight Instructor Certificate.

This guide was originally designed for use in a Part 141 school but has quickly become popular with those training under 14 CFR Part 61 who are not affiliated with an approved school. It will also prove beneficial to flight instructors who wish to refresh their knowledge or who are preparing to renew their flight instructor certificate.

The *Flight Instructor Practical Test Standards* book (FAA-S-8081-6AS) specifies the areas in which knowledge and skill must be demonstrated by the applicant before issuance of a flight instructor certificate with the associated category and class ratings. The *Certified Flight Instructor Oral Exam Guide* contains questions on the procedures and maneuvers in the Private Pilot, Commercial Pilot, and Instrument Rating Practical Test Standards. The performance standards for those procedures and maneuvers are also included; these standards should be mostly up to the skill level expected of the Commercial Pilot (FAA-S-8081- 12A), unless the maneuver only appears in the Private Pilot Practical Test Standards (FAA-8081-14S). In the latter case, the skill level expected of the flight instructor applicant is supposed to be "more precise" than that of a private pilot applicant, determined at the discretion of the examiner (FAA-S-8081-6AS).

During the exam, an FAA examiner will attempt to determine that the applicant is able to make a practical application of the fundamentals of instructing and is competent to teach the subject matter, procedures, and maneuvers included in the standards to students with varying backgrounds and levels of experience and ability. Through very intensive post-CFI-checkride de-briefings, we have provided you with the questions most consistently asked along with the information or the appropriate reference necessary for a knowledgeable response.

Continued

This guide may be supplemented with other comprehensive study materials as noted in parentheses after each question. For example: (H-8083-3). The abbreviations for these materials and their titles are listed below. Ensure that the latest revision of these references is used when reviewing for the test.

14 CFR Part 61	*Certification: Pilots, Flight Instructors, and Ground Instructors*
14 CFR Part 91	*General Operating and Flight Rules*
NTSB Part 830	*Notification and Reporting of Aircraft Accidents and Incidents*
AC 00-6	*Aviation Weather*
AC 00-45	*Aviation Weather Services*
H-8083-13	*Aviation Instructor Handbook*
H-8083-3	*Airplane Flying Handbook*
AC 61-23	*Pilot's Handbook of Aeronautical Knowledge*
AC 61-27	*Instrument Flying Handbook*
AC 61-65	*Certification of Pilots and Flight Instructors*
AC 61-67	*Use of Distractions During Flight Training*
AC 65-15	*Airframe & Powerplant Mechanics: Airframe Handbook*
AC 67-2	*Medical Handbook for Pilots*
H-8083-1	*Aircraft Weight and Balance Handbook*
AC 91-67	*Minimum Equipment Requirements for General Aviation Operations under Part 91*
AIM	*Aeronautical Information Manual*
FAA-S-8081-4	*Instrument Rating Practical Test Standards*
FAA-S-8081-6	*Flight Instructor Practical Test Standards*
FAA-S-8081-12	*Commercial Pilot Practical Test Standards*
FAA-S-8081-14	*Private Pilot Practical Test Standards*
POH	*Pilot Operating Handbook*
AFM	*FAA Approved Flight Manuals*

Most of the publications listed above are reprinted by ASA and are available from aviation retailers worldwide.

A review of the information presented within this guide should provide the necessary preparation for the FAA initial Certified Flight Instructor certification check.

Were you asked a question during your checkride that was not covered in this book? If so, please send the question to ASA. We are constantly striving to improve our publications to meet the industry needs.

e-mail: asa@asa2fly.com
Fax: 425.235.0128

7005 132nd Place SE
Newcastle, WA 98059-3153

Fundamentals of Instructing

1

A. The Learning Process

1. Briefly define the term "learning." (H-8083-13)

Learning can be defined as a change in behavior as a result of experience.

2. What are the basic characteristics of learning? (H-8083-13)

Learning is purposeful — Each student is a unique individual whose past experience affects readiness to learn and understanding of the requirements involved.

Learning comes through experience — Learning is an individual process. Knowledge cannot be poured into the student's head. The student can learn only from his/her individual experience.

Learning is multifaceted — The learning process may involve verbal, conceptual, perceptual, or emotional elements, and elements of problem solving all taking place at once.

Learning is an active process — For students to learn, they must react and respond.

3. What are the laws of learning? (H-8083-13)

The laws of learning are rules and principles that apply generally to the learning process. The first three are basic laws; the last three laws are the result of experimental studies:

R eadiness

E xercise

E ffect

P rimacy

I ntensity

R ecency

4. What is the "law of readiness"? (H-8083-13)

Individuals learn best when they are ready to learn, and they do not learn if they see no reason for learning. If students have a strong purpose, a clear objective, and a well-fixed reason for learning something, they make more progress than if they lack this kind of motivation.

5. What is the "law of exercise"? (H-8083-13)

The law of exercise states that those things most often repeated are best remembered. It is the basis of practice and drill.

6. What is the "law of effect"? (H-8083-13)

The law of effect states that learning is strengthened when accompanied by a pleasant or satisfying feeling, and that learning is weakened when associated with an unpleasant feeling.

7. What is the "law of primacy"? (H-8083-13)

Primacy, the state of being first, often creates a strong, almost unshakable, impression. What is taught must be right the first time.

8. What is the "law of intensity"? (H-8083-13)

A vivid, dramatic, or exciting learning experience teaches more than a routine or boring experience.

9. What is the "law of recency"? (H-8083-13)

The things most recently learned are best remembered.

10. What is the basis of all learning? (H-8083-13)

All learning comes from perceptions that are directed to the brain by one or more of the five senses (sight, hearing, touch, smell, and taste).

11. How do people learn? (H-8083-13)

All learning involves the following:

Perception — Initially all learning comes from perceptions which are directed to the brain by one or more of the five senses (sight, hearing, touch, smell and taste).

Insight — The grouping of perceptions into a meaningful whole.

Motivation — The most dominant force governing the student's progress and ability to learn.

12. What are the four levels of learning? (H-8083-13)

Rote learning — The ability to repeat back something which one has been taught, without understanding or being able to apply what has been learned.

Understanding — Perceiving and learning what has been taught.

Application — Achieving the skill to apply what has been learned and to perform correctly.

Correlation — Associating what has been learned with other things previously learned or encountered.

13. State several principles utilized in learning a skill. (H-8083-13)

a. *Physical skills involve more than muscles:* Perceptions change as the physical skill becomes easier.

b. *Desire to learn:* Shorter initial learning time and more rapid progress take place when a desire to learn exists.

c. *Patterns to follow:* The best way to prepare a student is to provide a clear step-by-step example.

d. *Perform the skill:* The student needs coordination between muscles and visual and tactile senses.

e. *Knowledge of results:* It is important for students to be aware of their progress.

Continued

f. *Progress follows a pattern:* Learning a skill usually follows a pattern. There is rapid improvement in the early stages, followed by a "leveling off" (commonly referred to as a "learning plateau").

g. *Duration and organization of the lesson:* In planning for student performance, a primary consideration is the length of time devoted to practice.

h. *Evaluation versus critique:* In the initial stages, practical suggestions are more valuable than a grade.

i. *Application of skill:* The student must use what has been learned. A student must learn the skill so well that it becomes easy to perform it.

14. Why do individuals forget what has been learned? (H-8083-13)

Disuse—A person forgets things which are not used.

Interference—People forget a thing because a certain experience has overshadowed it, or the learning of similar things has interfered.

Repression—Forgetting is due to the submersion of ideas into the unconscious mind. Individuals may unintentionally repress material that is unpleasant, or produces anxiety.

15. What actions can the instructor take to assist individuals in remembering what has been learned? (H-8083-13)

a. Praise stimulates remembering.

b. Recall is prompted by association.

c. Favorable attitudes aid retention—people learn and remember only what they wish to know.

d. Learning with all the senses is most effective—the best perception results from all senses working together.

e. Meaningful repetition aids recall.

16. Does positive or negative transfer of learning occur when a student learns a maneuver through learning a different maneuver? (H-8083-13)

This is positive transfer of learning.

17. What is the basic reason for use of the "building block" method of instruction? (H-8083-13)

The formation of correct habit patterns from the beginning of any learning process is essential to further learning and for correct performance after the completion of training. Each simple task is performed acceptably and correctly before the next learning task is introduced.

18. What are "defense mechanisms"? (H-8083-13)

Certain behavior patterns are called defense mechanisms because they are subconscious defenses against the realities of unpleasant situations.

19. What are four common defense mechanisms? (H-8083-13)

Rationalization—A subconscious technique for justifying actions that otherwise would be unacceptable; substitution of excuses for reasons.

Flight—Students often escape from frustrating situations by taking flight, physically or mentally; examples are faked illness or day-dreaming.

Aggression—Students may ask irrelevant questions, refuse to participate, etc., when they cannot deal directly with the cause of their frustration.

Resignation—Students may become so frustrated that they lose interest and give up.

Additional Study Questions:

1. **What factors will affect an individual's perceptions?** (H-8083-13)

2. **What are three elements of effective communication?** (H-8083-13)

3. **The effectiveness of the communicator is related to which three basic factors?** (H-8083-13)

4. **What are some of the barriers to effective communication?** (H-8083-13)

B. The Teaching Process

1. **What are the basic steps involved in the teaching process?** (H-8083-13)

   The teaching of new material can be broken down into the steps of:

   a. Preparation.

   b. Presentation.

   c. Application.

   d. Review and evaluation.

2. **What responsibilities does the flight instructor have in the "preparation" stage of a lesson?** (H-8083-13)

   For each lesson or instructional period, the instructor must determine:

   a. What is to be covered.

   b. The objectives of the lesson.

   c. The goals to be obtained.

   d. That all necessary supplies, materials, and equipment are readily available.

   e. That the equipment is operating properly.

3. **Should the instructor always use the same method for presentations?** (H-8083-13)

   No, depending on the material or information to be related, different methods of presentation can be used.

4. **Why is immediate application of knowledge beneficial to the student?** (H-8083-13)

   Good piloting habits can be established if a student uses the knowledge early in the learning process.

5. **What should the evaluation of a student's performance be based on?** (H-8083-13)

   The stated objectives and goals decided upon at the beginning of the lesson.

C. Teaching Methods

1. **What are the three main steps involved when organizing the material for a particular lesson?** (H-8083-13)

   a. Introduction
   b. Development
   c. Conclusion

2. **What basic elements should the introduction step contain?** (H-8083-13)

   Attention—Gain the student's attention and focus it on the subject involved.

   Motivation—Should appeal to each student personally and accentuate their desire to learn.

   Overview—Tell the student what is to be covered; give a clear, concise presentation of the objectives and key ideas; provide a road map of the route to be followed.

3. Discuss the development step of a presentation. (H-8083-13)

This is the main part of the lesson. The instructor develops the subject matter in a manner that helps the student achieve desired learning objectives. The instructor must logically organize the material to show the relationships of the main points.

4. Discuss the conclusion step of a presentation. (H-8083-13)

An effective conclusion retraces the important elements of the lesson and relates them to the objective. This review and wrap-up of ideas reinforces the student's learning and improves the retention of what has been learned.

5. What are the three most common teaching methods? (H-8083-13)

a. Lecture method

b. Guided discussion method

c. Demonstration/performance method

6. Discuss the lecture method of teaching. (H-8083-13)

The lecture is used primarily to introduce students to a new subject, but it is also a valuable method for summarizing ideas, showing relationships between theory and practice, and re-emphasizing main points.

7. What is the "guided discussion" method of teaching? (H-8083-13)

In contrast to the lecture method, where the instructor provides information, the guided discussion method relies on the students to provide ideas, experiences, opinions, and information. Through the skillful use of "lead-off" type questions, the instructor "draws out" what the student knows, rather than spending the class period telling them.

8. What are the different types of questions which may be used in a guided discussion? (H-8083-13)

Overhead—Directed at entire group; used to stimulate thought and response; good lead-off question.

Rhetorical—Used to stimulate thought; usually asked and answered by the instructor.

Direct—Used to get a response from a specific individual.

Reverse—Instructor redirects a student's question back in an effort to let the student provide the answer.

Relay—Instructor redirects student's question to the group for an answer instead of individual.

9. What is the demonstration/performance method of teaching? (H-8083-13)

This method of teaching is based on the simple yet sound principle that we learn by doing.

10. What are the five essential phases of the demonstration/performance method of teaching? (H-8083-13)

a. Explanation

b. Demonstration

c. Student performance

d. Instructor supervision

e. Evaluation

11. Define the term "programmed instruction." (H-8083-13)

This type of instruction has a student actively responding to each instructional step and receiving immediate feedback on their responses.

12. Define the term "integrated flight instruction."
(H-8083-13)

In integrated flight instruction, students are taught to perform flight maneuvers both by outside references and by reference to flight instruments, from the first time each maneuver is introduced.

Additional Study Questions:

1. What are the different ways an instructor may develop the main points of a lesson? (H-8083-13)

2. When presenting a particular subject or idea, what are some of the advantages and disadvantages associated with the lecture method? (H-8083-13)

3. When using the lecture method, what are some of the common types of presentations? (H-8083-13)

4. What is the basic objective of integrated flight instruction? (H-8083-13)

D. Evaluation

1. What is the basic purpose of a critique? (H-8083-13)

A critique should improve students' performance and provide them with something constructive with which to work and upon which they can build.

2. What are the characteristics of an effective critique?
(H-8083-13)

A critique should be:

a. *Objective*—Focused on student performance; should not reflect the personal opinions, etc., of the instructor.

b. *Flexible*—Fit in tone, technique, and content to the occasion and to the student.

c. *Acceptable*—Students must first accept the instructor. Effective critiques are presented with authority, conviction, sincerity, and from a position of recognizable competence.

d. *Comprehensive*—Cover a few major points or a few minor points, as well as the student's overall strengths and weaknesses.

e. *Constructive*—Provide positive guidance for correcting the faults and strengthening the weaknesses.

f. *Well-organized*—Follow some pattern of organization.

g. *Thoughtful*—Towards the student's need for self-esteem, recognition, and approval from others.

h. *Specific*—Comments and recommendations should not be so general that the student can find nothing to hold on to.

3. What types of oral questions should the instructor avoid? (H-8083-13)

a. "Puzzle" type questions with many parts and subparts.

b. "Oversize" questions that are too general, covering a wide area.

c. "Toss-up" questions where there is more than one correct answer.

d. "Bewilderment" questions that are not clear as to their content.

4. What should the instructor do and/or say if they do not know the answer to a question asked? (H-8083-13)

They should indicate to the student that they do not know the answer but will help the student find the correct answer.

5. What are the five characteristics of a good written test? (H-8083-13)

Reliability—it yields consistent results.

Validity—it measures what should be measured.

Useability—it is easy to understand and grade.

Comprehensiveness—it must sample whatever is being measured.

Discrimination—it will detect small differences.

6. Where can a flight instructor find the performance standards for evaluation of a particular maneuver? (H-8083-13)

The FAA publishes "Practical Test Standards" which establish the minimum standards for the certification of pilots. They are available from ASA and the GPO (Government Printing Office).

Additional Study Questions:

1. How can an instructor determine that a student is ready to solo? (H-8083-13)

2. What are some of the basic characteristics of an "effective" question? (H-8083-13)

3. What are several ways an instructor may evaluate a student's progress? (H-8083-13)

4. Why should an instructor use oral quizzing for evaluation of a student's performance? (H-8083-13)

5. What are two basic types of oral questions? (H-8083-13)

E. Flight Instructor Characteristics

1. **What major considerations and qualifications are included in the definition of professionalism?** (H-8083-13)

   a. Professionalism exists only when a service is performed for someone, or for the common good.

   b. Professionalism is achieved only after extended training and preparation.

   c. True performance as a professional is based on study and research.

   d. Professionalism presupposes an intellectual requirement. Professionals must be able to reason logically and accurately.

e. Professionalism requires the ability to make decisions with good judgment.

f. Professionalism demands a code of ethics. Professionals must be true to themselves, and to those they serve.

2. How would an instructor teach a student to counteract fear and anxiety? (H-8083-13)

Anxiety can be countered by reinforcing students' enjoyment of flying and by teaching them to cope with their fears. An effective technique is to treat fear as a normal reaction rather than ignoring it.

3. If an instructor has used personal contact and good advice to correct an observed pilot's unsafe operations and still cannot correct the situation, what should he/she do? (AC 60-41)

The instructor should report the observed deficiencies to an Accident Prevention Counselor or a GADO Accident Prevention Specialist.

4. Part 61 of the Federal Aviation Regulations covers rules for endorsements, but which Advisory Circular should the flight instructor be familiar with as well? (AC 61-65)

Advisory Circular 61-65, *Part 61 Certification: Pilot and Flight Instructors. See* the Appendix to review this document.

5. What guidelines should the instructor adhere to when determining whether a student is ready for the flight test? (AC 61-65)

The instructor should use the Practical Test Standards as a minimum standard for ensuring proper recommendation.

Additional Study Questions:

1. **What are several performance factors a professional flight instructor should attempt to achieve?** (H-8083-13)

2. **What is the single most significant psychological factor affecting student learning?** (H-8083-13)

3. **What are some examples of abnormal reactions to fear?** (H-8083-13)

F. Human Factors

1. **Control of human behavior involves understanding human needs. Name the five basic needs.** (H-8083-13)

   a. Physical

   b. Safety

   c. Social

   d. Egoistic

   e. Self-fulfillment

2. **How can an instructor develop a student's potential?** (H-8083-13)

   The instructor needs to view the student's potential as "vast and untapped." The instructor must discover what makes students "tick" and be an aid toward students' self-fulfillment.

3. **What rules should an instructor follow to ensure good human relations with the student?** (H-8083-13)

   a. Keep students motivated.

   b. Keep students informed.

   c. Approach students as individuals.

   d. Give credit when due.

   e. Criticize constructively.

   f. Be consistent.

   g. Admit errors.

✗ G. Planning Instructional Activity

1. What are the basic steps in planning a course of learning? (H-8083-13)

Before any important instruction can begin, the following must be considered:

a. Determination of standards and objectives.

b. Development and assembly of blocks of learning.

c. Identification of the blocks of learning.

2. What is a training syllabus? (H-8083-13)

A training syllabus consists of an outline of the course of training. Each block of learning is identified and presented in sequential order. The individual lessons within a block of learning will also be presented in sequential order.

3. What is a lesson plan? (H-8083-13)

A lesson plan is an organized outline or "blueprint" for a single instructional period and should be prepared in written form for each ground school and flight period. It should:

a. Tell what to do,

b. In what order to do it, and

c. What procedure to use in teaching it.

4. What items will a lesson plan always contain? (H-8083-13)

a. Lesson objective

b. Elements included

c. Schedule

d. Equipment

e. Instructor's actions

f. Student's actions

g. Completion standards

5. **What can an instructor do to make a lesson plan more flexible?** (H-8083-13)

The instructor can adapt or change the plan to accommodate the students and their different backgrounds, flight experience, and abilities.

Additional Study Questions:

1. **What items should each lesson contain within a training syllabus?** (H-8083-13)

2. **Why is use of a lesson plan important?** (H-8083-13)

3. **What are some of the basic characteristics of an effective lesson plan?** (H-8083-13)

4. **What are instructional aids and why are they utilized?** (H-8083-13)

H. Preflight Lesson on a Maneuver

An FAA examiner will determine that the applicant exhibits instructional knowledge of the elements related to the planning of instructional activity. This will be accomplished by requiring the applicant to develop a lesson plan for any one of the required maneuvers. The following is an example of a lesson plan for a 90-minute flight instruction period.

LESSON: Ground Reference Maneuvers

STUDENT:

DATE:

Maneuvers: Turns around a point, S-turns, Rectangular course

Objective: To develop the student's skill in planning and following a pattern over the ground compensating for wind drift at varying angles.

Elements: Use of ground references to control path; observation and control of wind effect; control of airplane attitude, altitude, and heading.

Schedule: Preflight discussion 10 minutes; Instructor demonstrations 25 minutes; Student practice 45 minutes; Postflight critique 10 minutes.

Equipment: Chalkboard for preflight discussion; IFR hood for review of previous maneuvers.

Instructor's Actions: Preflight— discuss lesson objective. Discuss Turns around a point, S-turns, Rectangular course; coach student practice.

Student's Actions: Preflight— discuss lesson objective. Resolve questions. Inflight—Review previous maneuvers including power-off stalls and slow flight. Perform each new maneuver as directed. Postflight—Ask pertinent questions.

Completion Standards: Student should demonstrate competency in maintaining orientation, airspeed within 10 knots, altitude within 100 feet, headings within 10 degrees, and in making proper correction for wind direction.

Flight Instructor Responsibilities

2

A. General

Note: Flight instructors should consult the most current edition of 14 CFR Part 61 for the *full* text of these regulations on pilot certification.

1. What is 14 CFR Part 61? (14 CFR 61.1)

Part 61 prescribes the requirements for issuing pilot, flight instructor, and ground instructor certificates and ratings, the conditions under which those certificates and ratings are necessary, and their associated privileges and limitations.

2. What are the various certificates issued under Part 61? (14 CFR 61.5)

a. Pilot certificates—student pilot, recreational pilot, private pilot, commercial pilot, airline transport pilot;

b. Flight instructor certificates; and

c. Ground instructor certificates.

3. What are the various ratings that may be placed on a pilot certificate (other than student pilot)? (14 CFR 61.5)

These ratings are placed on a pilot certificate (other than student pilot) when an applicant satisfactorily accomplishes the training and certification requirements for the rating sought:

a. Aircraft category ratings—Airplane, Rotorcraft, Glider, Lighter-than-air, Powered-lift.

b. Airplane class ratings—Single-engine land, Multi-engine land, Single-engine sea, Multi-engine sea.

c. Rotorcraft class ratings—Helicopter, Gyroplane.

d. Lighter-than-air class ratings—Airship, Balloon.

e. Aircraft type ratings—Large aircraft other than lighter-than-air, Turbojet-powered airplanes, Other aircraft type ratings specified through aircraft type certification procedures.

f. Instrument ratings (on private and commercial pilot certificates only)—Instrument Airplane, Instrument Helicopter, Instrument Powered-lift.

4. What are the various ratings that may be placed on a flight instructor certificate? (14 CFR 61.5)

These ratings are placed on a flight instructor certificate when an applicant satisfactorily accomplishes the training and certification requirements for the rating sought:

a. Aircraft category ratings—Airplane, Rotorcraft, Glider, Powered-lift.

b. Airplane class ratings—Single-engine, Multi-engine.

c. Rotorcraft class ratings—Helicopter, Gyroplane.

d. Instrument ratings—Instrument Airplane, Instrument Helicopter, Instrument Powered-lift.

5. A temporary pilot certificate that has been recently issued will remain effective for what length of time? (14 CFR 61.17)

120 days.

6. What is the duration of a flight instructor certificate? (14 CFR 61.19)

A flight instructor certificate is effective only while the holder has a current pilot certificate, and it expires 24 calendar months from the month in which it was issued or renewed.

7. What are the required medical certificates for the various pilot certificates? (14 CFR 61.23)

a. Must hold a first-class medical certificate when exercising the privileges of an airline transport pilot certificate;

b. Must hold at least a second-class medical certificate when exercising the privileges of a commercial pilot certificate; or

c. Must hold at least a third-class medical certificate when exercising the privileges of a private pilot certificate, a recreational pilot certificate, a student pilot certificate, a flight instructor certificate (except for a flight instructor certificate with a glider category rating, if the person is acting as the pilot-in-command or is serving as a required crewmember); or except for a glider category rating or a balloon class rating, prior to taking a practical test that is performed in an aircraft for a certificate or rating at the recreational, private, commercial, or airline transport pilot certificate level.

8. Give several examples of operations that do not require a medical certificate. (14 CFR 61.23)

a. When exercising the privileges of a flight instructor certificate if not acting as pilot-in-command or serving as a required crewmember;

b. When exercising the privileges of a ground instructor certificate;

c. When serving as an examiner or a check airman during the administration of a test or check for a certificate, rating, or authorization conducted in a flight simulator or flight training device; or

d. When taking a test or check for a certificate, rating, or authorization conducted in a flight simulator or flight training device.

9. In the event an airman certificate, medical certificate, or knowledge test report is lost or destroyed, what procedure should be followed? (14 CFR 61.29)

a. An application for the replacement of a lost or destroyed certificate or report is made by letter to the Department of Transportation, FAA;

b. A person who has lost a certificate or report may obtain a facsimile from the FAA confirming that it was issued. The facsimile may be carried as a certificate for a period not to exceed 60 days pending receipt of a duplicate certificate.

10. When is a type rating required? (14 CFR 61.31)

A type rating is required when acting as pilot-in-command of:

  a. Large aircraft (except lighter-than-air);

  b. Turbojet-powered airplanes;

  c. Other aircraft (specified through aircraft type certificate procedures).

11. According to regulations, what additional training is required to act as pilot-in-command of a complex airplane? (14 CFR 61.31)

No person may act as pilot-in-command of a complex airplane (an airplane with retractable landing gear, flaps, and controllable-pitch propeller, or in the case of a seaplane, flaps, and controllable-pitch propeller), unless the person has:

  a. Received and logged ground and flight training from an authorized instructor in a complex airplane, or in a flight simulator or flight training device (representative of a complex airplane), and has been found proficient in its operation and systems;

  b. Received a one-time logbook endorsement from an authorized instructor certifying the person is proficient to operate a complex airplane.

Note: This is not required if the person has logged flight time as pilot-in-command of a complex airplane, or in a flight simulator or flight training device prior to August 4, 1997.

12. According to regulations, what additional training is required to act as pilot-in-command of a high-performance airplane? (14 CFR 61.31)

No person may act as pilot-in-command of a high-performance airplane (an engine of more than 200 horsepower), unless the person has:

  a. Received and logged ground and flight training from an authorized instructor in a high-performance airplane, or in a flight simulator or flight training device (representative of a high-performance airplane), and has been found proficient in its operation and systems;

b. Received a one-time logbook endorsement from an authorized instructor certifying the person is proficient to operate a high-performance airplane.

Note: This is not required if the person has logged flight time as pilot-in-command of a high-performance airplane, or in a flight simulator or flight training device representative of a high-performance airplane prior to August 4, 1997.

13. Describe the minimum required elements of instruction for transition to a tailwheel aircraft. (14 CFR 61.31)

No person may act as pilot-in-command of a tailwheel airplane unless that person has received and logged flight training from an authorized instructor in a tailwheel airplane, and has received a logbook endorsement from an authorized instructor who found the person proficient in the operation of a tailwheel airplane, to include at least these maneuvers and procedures: normal and crosswind takeoffs and landings, wheel landings (unless the manufacturer has recommended against such landings), and go-around procedures. This is not required if the person logged pilot-in-command time in a tailwheel airplane before April 15, 1991.

14. To be eligible to take an FAA knowledge test, what must an applicant accomplish? (14 CFR 61.35)

An applicant must have received an endorsement from an authorized instructor certifying that the applicant accomplished a required ground-training or a home-study course for the certificate or rating sought, and is prepared for the knowledge test; and must have proper identification at the time of application that contains the applicant's photograph, signature, date of birth (which shows the applicant meets or will meet the age requirements of this part for the certificate sought before the expiration date of the airman knowledge test report), and actual residential address (if different from the applicant's mailing address).

15. What various methods may be used by a student to show evidence of ground school or home study course completion? (AC 61-65)

a. A certificate of graduation from a pilot training course by an FAA-certificated pilot school.

b. An endorsement from an appropriately rated FAA-certificated ground or flight instructor.

c. A certificate of graduation or statement of accomplishment from a ground school course (high school, college, adult education program, etc.).

d. A certificate of graduation from an industry-provided aviation home study course.

e. An endorsement from an appropriately rated FAA-certificated ground or flight instructor for completion of an individually developed home study course.

16. What actions is an applicant required to have accomplished in order to be eligible for a practical test for a certificate or rating issued under 14 CFR Part 61? (14 CFR 61.39)

a. Pass the required knowledge test within the 24-calendar-month period preceding the month the applicant completes the practical test, if a knowledge test is required;

b. Present the knowledge test report at the time of application for the practical test, if a knowledge test is required;

c. Have satisfactorily accomplished the required training and obtained the aeronautical experience prescribed by this part for the certificate or rating sought;

d. Hold at least a current third-class medical certificate, if a medical certificate is required;

e. Meet the prescribed age requirement of this part for the issuance of the certificate or rating sought;

f. Have a logbook or training record endorsement, if required by this part, signed by an authorized instructor who certifies that the applicant: has received and logged training time within 60 days preceding the date of application in preparation for the practical test; is prepared for the required practical test; and has demonstrated satisfactory knowledge of subject areas in which the applicant was deficient on the knowledge test; have a completed and signed application form.

17. Other than the Practical Test Standards, what general guidelines will an examiner follow when judging the ability of an applicant for a pilot certificate or rating? (14 CFR 61.43)

The examiner will judge based on that applicant's ability to safely:

a. Perform the tasks specified in the areas of operation for the certificate or rating sought within approved standards;

b. Demonstrate mastery of the aircraft with the successful outcome of each task performed never seriously in doubt;

c. Demonstrate satisfactory proficiency and competency within approved standards;

d. Demonstrate sound judgment; and

e. Demonstrate single-pilot competence if the aircraft is type certificated for single-pilot operations.

18. An applicant for a knowledge or practical test who fails that test may only reapply when what actions have been accomplished? (14 CFR 61.49)

An applicant who fails may only reapply for the test after the applicant has received:

a. The necessary training from an authorized instructor, who has determined that the applicant is proficient to pass the test; and

b. An endorsement from an authorized instructor who gave the applicant the additional training.

19. What type of flight time must be documented and recorded by all pilots? (14 CFR 61.51)

a. Training and aeronautical experience used to meet the requirements for a certificate, rating, or flight review.

b. The aeronautical experience required for meeting the recency of flight experience requirements of Part 61.

20. What regulatory guidelines are established concerning the logging of instrument time? (14 CFR 61.51)

a. A person may log instrument flight time only for that flight time when the person operates the aircraft solely by reference to instruments under actual or simulated instrument flight conditions.

b. An authorized instructor may log instrument flight time when conducting instrument flight instruction in actual instrument flight conditions.

c. For the purposes of logging instrument flight time to meet the recent instrument experience requirements of §61.57(c) the following information must be recorded in the person's logbook: the location and type of each instrument approach accomplished; and the name of the safety pilot, if required.

d. A flight simulator or flight training device may be used by a person to log instrument flight time, provided an authorized instructor is present during the simulated flight.

21. When entering time in a logbook, what time is considered as "training time"? (14 CFR 61.51)

a. A person may log training time when that person receives training from an authorized instructor in an aircraft, flight simulator, or flight training device.

b. The training time must be logged in a logbook and must be endorsed in a legible manner by the authorized instructor; and include a description of the training given, the length of the training lesson, and the instructor's signature, certificate number, and certificate expiration date.

22. What requirements must be met when conducting a Biennial Flight Review? (14 CFR 61.56)

A flight review consists of a minimum of 1 hour of flight training and 1 hour of ground training, including a review of the current general operating and flight rules of Part 91 and a review of those maneuvers and procedures that, at the discretion of the person giving the review, are necessary for the pilot to demonstrate the safe exercise of the privileges of the pilot certificate.

23. What logbook entries should be made by the flight instructor if a pilot demonstrates unsatisfactory performance during a Biennial Flight Review? (AC 61-65)

The instructor should sign the logbook to record the instruction given, and then recommend additional training in the areas of the review that were unsatisfactory. The instructor should not make record of an unsatisfactory review in the logbook.

24. When is a flight review not required? (14 CFR 61.56)

Pilots who have satisfactorily completed within the preceding 24 calendar months before the month in which they act as pilot-in-command:

a. A pilot proficiency check conducted by the FAA, an approved pilot check airman, or U.S. Armed Forces, for a pilot certificate, rating, or operating privilege.

b. Completed one or more phases of an FAA-sponsored pilot proficiency award program.

Also, a flight instructor who holds a current flight instructor certificate and has satisfactorily completed renewal of a flight instructor certificate need not accomplish the 1 hour of ground instruction.

Note: The flight review may be accomplished in combination with the requirements of §61.57 and other applicable recency-of-experience requirements at the discretion of the instructor.

Continued

25. What are the "recency-of-experience" requirements for acting as pilot-in-command of an aircraft carrying passengers? (14 CFR 61.57)

No person may act as a pilot-in-command of an aircraft carrying passengers, or as a required pilot onboard an aircraft that requires more than one pilot flight crewmember, unless that person has made at least three takeoffs and three landings within the preceding 90 days, and:

a. The person acted as the sole manipulator of the flight controls;

b. The required takeoffs and landings were performed in an aircraft of the same category, class, and type (if a type rating is required), and if the aircraft to be flown is an airplane with a tailwheel, the takeoffs and landings must have been made to a full stop in an airplane with a tailwheel.

For the purpose of meeting these requirements, a person may act as a pilot-in-command of an aircraft under day VFR or day IFR, provided no persons or property are carried onboard the aircraft, other than those necessary for the conduct of the flight.

26. When acting as PIC of an aircraft carrying passengers at night, what recent night experience must the pilot have accomplished? (14 CFR 61.57)

No person may act as pilot-in-command of an aircraft carrying passengers during the period beginning 1 hour after sunset and ending 1 hour before sunrise, unless within the preceding 90 days that person:

a. Has made at least three takeoffs and three landings to a full stop during the period beginning 1 hour after sunset and ending 1 hour before sunrise;

b. Has acted as sole manipulator of the flight controls; and

c. Has performed the required takeoffs and landings in an aircraft of the same category, class and type (if a type rating is required).

B. Aircraft Rating and Special Certification

1. If a pilot wants to add an additional category rating to his or her pilot certificate, what is required?
(14 CFR 61.63)

This person must:

a. Have received the required training and possess the prescribed aeronautical experience that applies to the pilot certificate for the aircraft category and, if applicable, class rating sought;

b. Have an endorsement in his or her logbook or training record from an authorized instructor, attesting that the applicant has been found competent in the appropriate aeronautical knowledge areas, and has been found proficient in the areas of operation appropriate to the pilot certificate for the aircraft category and, if applicable, class rating sought;

c. Pass the required practical test appropriate to the pilot certificate for the aircraft category and, if applicable, class rating sought.

This person need not take an additional knowledge test, provided that he or she holds an airplane, rotorcraft, powered-lift, or airship rating at that pilot certificate level.

2. If a pilot wants to add on an additional class rating to his or her pilot certificate, what is required? (14 CFR 61.63)

This person must:

a. Have an endorsement in his or her logbook or training record from an authorized instructor and that endorsement must attest that the applicant has been found competent in the aeronautical knowledge areas and proficient in the areas of operation, appropriate to the pilot certificate for the aircraft class rating sought;

b. Must pass the required practical test that is appropriate to the pilot certificate for the aircraft class rating sought.

This person need not meet the specified training time requirements that apply to the pilot certificate for the aircraft class rating sought, nor take an additional knowledge test, provided he or she holds an airplane, rotorcraft, powered-lift, or airship rating at that pilot certificate level.

3. **For an applicant training under Part 61, what minimum aeronautical experience, is required before application for an instrument rating may take place?** (14 CFR 61.65)

A person who applies for an instrument rating must have logged the following:

a. At least 50 hours of cross-country flight time as pilot-in-command, of which at least 10 hours must be in airplanes for an Instrument Airplane rating; and

b. A total of 40 hours of actual or simulated instrument time on the Part 61 areas of operation, to include at least 15 hours of instrument flight training from an authorized instructor in the aircraft category for which the instrument rating is sought; at least 3 hours of instrument training that is appropriate to the instrument rating sought from an authorized instructor (in preparation for the practical test within the 60 days preceding the date of the test); and for an Instrument Airplane rating, instrument training on cross-country flight procedures specific to airplanes that includes at least one cross-country flight in an airplane under IFR, and consists of a distance of at least 250 NM along airways or ATC-directed routing, an instrument approach at each airport, and three different kinds of approaches with the use of navigation systems.

4. **In what minimum aeronautical knowledge areas must a person who applies for an instrument rating have received instruction from an authorized instructor or home-study course?** (14 CFR 61.65)

a. Federal Aviation Regulations that apply to flight operations under IFR.

b. Appropriate information from the AIM that applies to flight operations under IFR.

c. Air traffic control system and procedures for instrument flight operations.

d. IFR navigation and approaches by use of navigation systems.

e. Use of IFR enroute and instrument approach procedure charts.

   f. Procurement and use of aviation weather reports and forecasts and the elements of forecasting weather trends based on that information and personal observation of weather conditions.

   g. Safe and efficient operation of aircraft under instrument flight rules and conditions.

   h. Recognition of critical weather situations and windshear avoidance.

   i. Aeronautical decision-making and judgment.

   j. Crew resource management, including crew communication and coordination.

5. To be eligible for an instrument rating, in what minimum areas of operation must the applicant have received training? (14 CFR 61.65)

A person who applies for an instrument rating must receive and log training from an authorized instructor in an aircraft, or in an approved flight simulator or approved flight training device that includes the following areas of operation:

   a. Preflight preparation;

   b. Preflight procedures;

   c. Air traffic control clearances and procedures;

   d. Flight by reference to instruments;

   e. Navigation systems;

   f. Instrument approach procedures;

   g. Emergency operations; and

   h. Postflight procedures.

C. Student Pilots

1. What are the eligibility requirements for a student pilot certificate? (14 CFR 61.83)

An applicant must:

a. Be at least 16 years of age for other than the operation of a glider or balloon.

b. Be at least 14 years of age for the operation of a glider or balloon.

c. Be able to read, speak, write, and understand the English language.

If the applicant is unable to meet one of these requirements due to medical reasons, then the Administrator may place such operating limitations on that applicant's pilot certificate as are necessary for the safe operation of the aircraft.

2. How can a person obtain a student pilot certificate? (14 CFR 61.85)

An application for a student pilot certificate is made on a form and in a manner provided by the Administrator and is submitted to:

a. A designated aviation medical examiner if applying for an FAA medical certificate under 14 CFR Part 67;

b. An examiner; or

c. A Flight Standards District Office.

3. What minimum aeronautical knowledge must be demonstrated by a student pilot before solo privileges are permitted? (14 CFR 61.87)

A student pilot must demonstrate satisfactory aeronautical knowledge on a knowledge test.

a. The test must address the student pilot's knowledge of:

i. Applicable sections of 14 CFR Parts 61 and 91;

ii. Airspace rules and procedures for the airport where the solo flight will be performed; and

   iii. Flight characteristics and operational limitations for the make and model of aircraft to be flown.

b. The student's authorized instructor must:

   i. Administer the test; and

   ii. At the conclusion of the test, review all incorrect answers with the student before authorizing that student to conduct a solo flight.

4. What minimum flight training must a student pilot receive before solo privileges are permitted?
(14 CFR 61.87)

Prior to conducting a solo flight, a student pilot must have:

a. Received and logged flight training for the Part 61 maneuvers and procedures appropriate to the make and model of aircraft to be flown; and

b. Demonstrated satisfactory proficiency and safety, as judged by an authorized instructor, on the Part 61 maneuvers and procedures in the make and model of aircraft or similar make and model of aircraft to be flown.

A student pilot training for a single-engine airplane rating must receive and log flight training for the following maneuvers and procedures:

a. Proper flight preparation procedures, including preflight planning and preparation, powerplant operation, and aircraft systems;

b. Taxiing or surface operations, including runups;

c. Takeoffs and landings, including normal and crosswind;

d. Straight-and-level flight, and turns in both directions;

e. Climbs and climbing turns;

f. Airport traffic patterns, including entry and departure procedures;

g. Collision avoidance, windshear avoidance, and wake turbulence avoidance;

h. Descents, with and without turns, using high and low drag configurations;

Continued

i. Flight at various airspeeds from cruise to slow flight;

j. Stall entries from various flight attitudes and power combinations with recovery initiated at the first indication of a stall, and recovery from a full stall;

k. Emergency procedures and equipment malfunctions;

l. Ground reference maneuvers;

m. Approaches to a landing area with simulated engine malfunctions;

n. Slips to a landing; and

o. Go-arounds.

5. What basic requirements must a student pilot meet before being allowed to conduct solo flight at night? (14 CFR 61.87)

a. Flight training at night on night-flying procedures, including takeoffs, approaches, landings, and go-arounds at night at the airport where the solo flight will be conducted;

b. Navigation training at night in the vicinity of the airport where the solo flight will be conducted;

c. An endorsement in the student's logbook for the specific make and model aircraft to be flown for night solo flight, by an authorized instructor who gave the training within the 90-day period preceding the date of the flight.

6. What limitations are imposed upon student pilots operating an aircraft in solo flight? (14 CFR 61.87)

A student pilot may not operate an aircraft in solo flight unless they received:

a. An endorsement from an authorized instructor on his or her student pilot certificate for the specific make and model aircraft to be flown; and

b. An endorsement in the student's logbook for the specific make and model aircraft to be flown by an authorized instructor, who gave the training within the 90 days preceding the date of the flight.

7. What limitations are imposed upon flight instructors authorizing student pilot solo flights? (14 CFR 61.87)

No instructor may authorize a student pilot to perform a solo flight unless that instructor has:

a. Given that student pilot training in the make and model of aircraft or a similar make and model of aircraft in which the solo flight is to be flown;

b. Determined the student pilot is proficient in the Part 61 prescribed maneuvers and procedures;

c. Determined the student pilot is proficient in the make and model of aircraft to be flown;

d. Ensured that the student pilot's certificate has been endorsed by an instructor authorized to provide flight training for the specific make and model aircraft to be flown; and

e. Endorsed the student pilot's logbook for the specific make and model aircraft to be flown, and that endorsement remains current for solo flight privileges, provided an authorized instructor updates the student's logbook every 90 days thereafter.

The flight training required by this section must be given by an instructor authorized to provide flight training who is appropriately rated and current.

8. State the general limitations which apply to all student pilots. (14 CFR 61.89)

A student pilot may not act as pilot-in-command of an aircraft:

a. Carrying a passenger;

b. Carrying property for compensation or hire;

c. For compensation or hire;

d. In furtherance of a business;

e. On an international flight, (with exceptions — see §61.89)

f. With a flight or surface visibility of less than 3 statute miles during daylight hours or 5 statute miles at night;

g. When the flight cannot be made with visual reference to the surface; or

Continued

h. In a manner contrary to any limitations placed in the pilot's logbook by an authorized instructor.

A student pilot may not act as a required pilot flight crewmember on any aircraft for which more than one pilot is required by the type certificate of the aircraft or regulations under which the flight is conducted, except when receiving flight training from an authorized instructor onboard an airship, and no person other than a required flight crewmember is carried on the aircraft.

9. **What requirements must be met before a flight instructor can allow a student pilot to make repeated specific solo cross-country flights without each flight being logbook endorsed?** (14 CFR 61.93)

Repeated specific solo cross-country flights may be made to another airport within 50 NM of the airport from which the flight originated, provided:

a. The authorized instructor has given the student flight training in both directions over the route, including entering and exiting the traffic patterns, takeoffs, and landings at the airports to be used;

b. The authorized instructor who gave the training has endorsed the student's logbook certifying the student is proficient to make such flights;

c. The student has current solo flight endorsements in accordance with §61.87; and

d. The student has current solo cross-country flight endorsements in accordance with §61.87(c); however, for repeated solo cross-country flights to another airport within 50 NM from which the flight originated, separate endorsements are not required to be made for each flight.

10. Before a student pilot is permitted solo cross-country privileges, they must have received several endorsements. What are they? (14 CFR 61.93)

Student pilot certificate endorsement:

A student pilot must have a solo cross-country endorsement from the authorized instructor who conducted the training, and that endorsement must be placed on that person's student pilot certificate for the specific category of aircraft to be flown.

Logbook endorsement:

a. A student pilot must have a solo cross-country endorsement from an authorized instructor placed in the student pilot's logbook for the specific make and model of aircraft to be flown.

b. A certificated pilot who is receiving training for an additional aircraft category or class rating must have an endorsement in his or her logbook or training record from an authorized instructor, and that endorsement must attest that the applicant has been found competent in the areas of aeronautical knowledge and operation that are appropriate to the pilot certificate for the aircraft category and, if applicable, class rating sought.

c. For each cross-country flight, the authorized instructor who reviews the cross-country planning must make an endorsement in the person's logbook after reviewing that person's cross-country planning, as specified in §61.93(d). The endorsement must specify the make and model of aircraft to be flown, state that the student's preflight planning and preparation is correct and that the student is prepared to make the flight safely under the known conditions, and state that any limitations required by the student's instructor are met.

11. Before solo cross-country privileges are permitted, what minimum cross-country flight training requirements must a student pilot satisfy? (14 CFR 61.93)

The student pilot must receive and log flight training in the following maneuvers and procedures:

a. Aeronautical charts for VFR navigation using pilotage and dead reckoning with aid of a magnetic compass;

b. Use of aircraft performance charts for cross-country flight;

c. Procurement and analysis of aeronautical weather reports and forecasts, recognizing critical weather situations and estimating visibility while in flight;

d. Emergency procedures;

e. Traffic pattern procedures that include area departure, area arrival, entry into the traffic pattern, and approach;

f. Collision avoidance, wake turbulence precautions, and windshear avoidance;

g. Recognition, avoidance, and operational restrictions of hazardous terrain features in the geographical area where the cross-country flight will be flown;

h. Proper operation of the instruments and equipment installed in the aircraft to be flown;

i. Use of radios for VFR navigation and two-way communications;

j. Takeoff, approach, and landing procedures, including short-field, soft-field, and crosswind takeoffs, approaches, and landings;

k. Climbs at best angle and best rate; and

l. Control and maneuvering solely by reference to flight instruments, including straight-and-level flight, turns, descents, climbs, use of radio aids, and ATC directives.

12. Before a flight instructor can authorize a student pilot to conduct a solo cross-country flight, what requirements must be met? (14 CFR 61.93)

The instructor must have determined that:

a. The student's cross-country planning is correct for the flight;

b. Upon review, the current and forecast weather conditions show that the flight can be completed under VFR;

c. The student is proficient to conduct the flight safely; and

d. The student has the appropriate and current solo cross-country endorsement for the make and model of aircraft to be flown;

e. The student's solo flight endorsement is current for the make and model aircraft to be flown.

13. What actions must a flight instructor take to allow a student pilot to operate within Class B airspace or at airports within Class B airspace? (14 CFR 61.95)

A student pilot may not operate an aircraft on a solo flight in Class B airspace unless:

a. The student pilot has received both ground and flight training from an authorized instructor on that Class B airspace area, and the flight training was received in the specific Class B airspace area for which solo flight is authorized;

b. The logbook of that student pilot has been endorsed by the authorized instructor who gave the student pilot flight training, and the endorsement is dated within the 90-day period preceding the date of the flight in that Class B airspace area; and

c. The logbook endorsement specifies that the student pilot has received the required ground and flight training, and has been found proficient to conduct solo flight in that specific Class B airspace area.

D. Recreational Pilots

1. What are the eligibility requirements for a recreational pilot certificate? (14 CFR 61.96)

a. Be at least 17 years of age;

b. Be able to read, speak, write, and understand the English language;

c. Receive a logbook endorsement from an authorized instructor who conducted the training or reviewed the applicant's home study on the required aeronautical knowledge areas, and certified that the applicant is prepared for the required knowledge test;

d. Pass the required knowledge test on the aeronautical knowledge areas;

e. Receive flight training and a logbook endorsement from an authorized instructor who conducted the training on the required areas of operation, and certified that the applicant is prepared for the required practical test;

f. Meet the required aeronautical experience;

g. Pass the required practical test on the areas of operation that apply to the aircraft category and class rating sought.

2. What aeronautical experience must an applicant for a recreational pilot certificate have accomplished? (14 CFR 61.99)

An applicant must receive and log at least 30 hours of flight training time that includes at least:

a. 15 hours of flight training from an authorized instructor on the required areas of operation that consists of at least 2 hours of flight training en route to an airport more than 25 NM from the airport where the applicant normally trains, which includes at least three takeoffs and three landings at the airport, and 3 hours of flight training in the aircraft for the rating sought in preparation for the practical test within the 60 days preceding the date of the practical test; and

b. 3 hours of solo flying in the aircraft for the rating sought, on the areas of operation listed in §61.98 of this part that apply to the aircraft category and class rating sought.

3. What are the privileges that apply to recreational pilots? (14 CFR 61.101)

A recreational pilot may:

a. Carry no more than one passenger; and

b. Not pay less than the pro rata share of the operating expenses of a flight with a passenger, provided the expenses involve only fuel, oil, airport expenses, or aircraft rental fees.

c. Act as pilot-in-command of an aircraft on a flight within 50 NM from the departure airport, provided that person has:

   i. Received ground and flight training for takeoff, departure, arrival, and landing procedures at the departure airport;

   ii. Received ground and flight training for the area, terrain, and aids to navigation that are in the vicinity of the departure airport;

   iii. Been found proficient to operate the aircraft at the departure airport and the area within 50 NM from that airport; and

   iv. Received from an authorized instructor a logbook endorsement, which is carried in the person's possession in the aircraft, that permits flight within 50 NM from the departure airport.

d. Act as pilot-in-command of an aircraft on a flight that exceeds 50 NM from the departure airport, provided that person has:

   i. Received ground and flight training from an authorized instructor on the required cross-country training that applies to the aircraft rating held;

   ii. Been found proficient in cross-country flying; and

   iii. Received from an authorized instructor a logbook endorsement, which is in the person's possession in the aircraft, that certifies the person has received and been found proficient in the required cross-country training that applies to the aircraft rating held.

4. What are the limitations which apply to all recreational pilots? (14 CFR 61.101)

A recreational pilot may not act as pilot-in-command of an aircraft:

a. That is certificated for more than four occupants, with more than one powerplant, with a powerplant of more than 180 horsepower, or with retractable landing gear.

b. That is classified as a multi-engine airplane, powered-lift, glider, airship, or balloon;

c. That is carrying a passenger or property for compensation or hire;

d. For compensation or hire;

e. In furtherance of a business;

f. Between sunset and sunrise;

g. In airspace in which communication with air traffic control is required;

h. At an altitude of more than 10,000 feet MSL or 2,000 feet AGL, whichever is higher;

i. When the flight or surface visibility is less than 3 statute miles;

j. Without visual reference to the surface;

k. On a flight outside the United States;

l. To demonstrate that aircraft in flight to a prospective buyer;

m. That is used in a passenger-carrying airlift and sponsored by a charitable organization; and

n. That is towing any object.

A recreational pilot may not act as a pilot flight crewmember on any aircraft for which more than one pilot is required by the type certificate of the aircraft or the regulations under which the flight is conducted, except when receiving flight training from a person authorized to provide flight training onboard an airship, and no person other than a required flight crewmember is carried on the aircraft.

E. Private Pilots

1. What are the general eligibility requirements for a private pilot certificate? (14 CFR 61.103)

To be eligible for a private pilot certificate, a person must:

a. Be at least 17 years of age for a rating in other than a glider or balloon;

b. Be at least 16 years of age for a rating in a glider or balloon;

c. Be able to read, speak, write, and understand the English language;

d. Receive a logbook endorsement from an authorized instructor who conducted the training or reviewed the person's home study on the required aeronautical knowledge areas and certified that the person is prepared for the required knowledge test;

e. Pass the required knowledge test on the aeronautical knowledge areas;

f. Receive flight training and a logbook endorsement from an authorized instructor who conducted the training in the required areas of operation and certified that the person is prepared for the required practical test;

g. Meet the aeronautical experience requirements that apply to the aircraft rating sought before applying for the practical test;

h. Pass a practical test on the required areas of operation that apply to the aircraft rating sought.

2. In what areas of aeronautical knowledge must an applicant for a private pilot certificate have received instruction? (14 CFR 61.105)

a. Federal Aviation Regulations that relate to private pilot privileges, limitations, and flight operations;

b. NTSB accident reporting requirements;

c. Use of the *Aeronautical Information Manual* and FAA ACs;

d. Aeronautical charts for VFR navigation using pilotage, dead reckoning, and navigation systems;

Continued

e. Radio communication procedures;

f. Recognition of critical weather situations from the ground and in flight, windshear avoidance, and the procurement and use of aeronautical weather reports and forecasts;

g. Safe and efficient operation of aircraft, including collision avoidance, and recognition and avoidance of wake turbulence;

h. Effects of density altitude on takeoff and climb performance;

i. Weight and balance computations;

j. Principles of aerodynamics, powerplants, and aircraft systems;

k. Stall awareness, spin entry, spins, and spin recovery techniques for the airplane and glider category ratings;

l. Aeronautical decision making and judgment; and

m. Preflight action that includes how to obtain information on runway lengths at airports of intended use, data on takeoff and landing distances, weather reports and forecasts, and fuel requirements, and how to plan for alternatives if the planned flight cannot be completed or delays are encountered.

3. In what pilot operations must an applicant for a private pilot certificate have received instruction?
(14 CFR 61.107)

a. Preflight preparation

b. Preflight procedures

c. Airport and seaplane base operations

d. Takeoffs, landings, and go-arounds

e. Performance maneuvers

f. Ground reference maneuvers

g. Navigation

h. Slow flight and stalls

i. Basic instrument maneuvers

j. Emergency operations

k. Night operations

l. Postflight procedures

4. What minimum aeronautical experience must be accumulated by a student pilot before application for a private pilot certificate? (14 CFR 61.109)

Total Time: 40 hours of flight time which consists of at least—

Dual: 20 hours of flight training with an instructor on the Private Pilot areas of operation that includes:

a. 3 hours of cross-country flight training in a single-engine airplane;

b. 3 hours of night flight training in a single-engine airplane, that includes at least:

    i. 1 cross-country flight of over 100 NM total distance; and

    ii. 10 takeoffs and 10 landings, each involving a flight in the traffic pattern.

c. 3 hours of flight training by reference to instruments in a single-engine airplane; and

d. 3 hours of flight training in a single-engine airplane within the preceding 60 days prior to the practical test.

Solo: 10 hours of solo flying in a single-engine airplane on the Private Pilot areas of operation, that includes:

a. 5 hours of solo cross-country flying;

b. 1 solo cross-country flight of at least 150 NM total distance with full stop landings at a minimum of 3 points and one segment of at least 50 NM between takeoff and landing; and

c. 3 takeoffs and landings to full stop (with each landing involving a flight in the traffic pattern) at a controlled airport.

5. May a student pilot take the private pilot practical test without completing the 3 hours of night flying instruction? (14 CFR 61.110)

No, with one exception: student pilots who receive flight training in and reside in the state of Alaska are not required to comply with the night flight training requirements.

6. What privileges and limitations apply to private pilots? (14 CFR 61.113)

a. No person who holds a private pilot certificate may act as pilot-in-command of an aircraft carrying passengers or property for compensation or hire.

b. A private pilot may, for compensation or hire, act as pilot-in-command of an aircraft in connection with any business or employment if the flight is only incidental to that business or employment, and the aircraft does not carry passengers or property for compensation or hire.

c. A private pilot may not pay less than the pro rata share of operating expenses of a flight with passengers, provided the expenses involve only fuel, oil, airport expenditures, or rental fees.

d. A private pilot may act as pilot-in-command of an aircraft used in a passenger-carrying airlift sponsored by a charitable organization described in 14 CFR §61.113(d)(7) for which the passengers make a donation to the organization, when the requirements of §61.113(d) are met.

e. A private pilot may be reimbursed for aircraft operating expenses directly related to search and location operations, provided the expenses involve only fuel, oil, airport expenditures, or rental fees, and the operation is sanctioned and under the direction and control of a local, State, or Federal agency, or an organization that conducts search and location operations.

f. A private pilot who is an aircraft salesman and who has at least 200 hours of logged flight time may demonstrate an aircraft in flight to a prospective buyer.

g. A private pilot who meets the requirements of §61.69 may act as pilot-in-command of an aircraft towing a glider.

F. Commercial Pilots

1. What are the general eligibility requirements for a commercial pilot certificate? (14 CFR 61.123)

a. Be at least 18 years of age;

b. Be able to read, speak, write, and understand the English language;

c. Receive a logbook endorsement from an authorized instructor who conducted the required ground training or reviewed the person's home study on the required aeronautical knowledge areas that apply to the aircraft category and class rating sought, and certified that the person is prepared for the required knowledge test;

d. Pass the required knowledge test on the aeronautical knowledge areas;

e. Receive the required training and a logbook endorsement from an authorized instructor who conducted the training on the required areas of operation that apply to the aircraft category and class rating sought, and certified that the person is prepared for the required practical test;

f. Meet the required aeronautical experience before applying for the practical test;

g. Pass the required practical test on the required areas of operation;

h. Hold at least a private pilot certificate; and

i. Comply with the sections of Part 61 that apply to the aircraft category and class rating sought.

2. In what areas of aeronautical knowledge must an applicant for a commercial pilot certificate have received instruction? (14 CFR 61.125)

a. Federal Aviation Regulations that relate to commercial pilot privileges, limitations, and flight operations;

b. NTSB accident reporting requirements;

c. Basic aerodynamics and the principles of flight;

Continued

d. Meteorology (recognition of critical weather situations, windshear recognition and avoidance, and the use of aeronautical weather reports and forecasts);

e. Safe and efficient operation of aircraft;

f. Weight and balance computations;

g. Use of performance charts;

h. Significance and effects of exceeding aircraft performance limitations;

i. Use of aeronautical charts and a magnetic compass for pilotage and dead reckoning;

j. Use of air navigation facilities;

k. Aeronautical decision making and judgment;

l. Principles and functions of aircraft systems;

m. Maneuvers, procedures, and emergency operations appropriate to the aircraft;

n. Night and high-altitude operations;

o. Procedures for operating within the National Airspace System; and

p. Procedures for flight and ground training for lighter-than-air ratings.

3. In what pilot operations must an applicant for a commercial pilot certificate have received instruction? (14 CFR 61.127)

a. Preflight preparation

b. Preflight procedures

c. Airport and seaplane base operations

d. Takeoffs, landings, and go-arounds

e. Performance maneuvers

f. Ground reference maneuvers

g. Navigation

h. Slow flight and stalls

   i. Emergency operations

   j. High-altitude operations

   k. Postflight procedures

4. What minimum aeronautical experience must be accumulated by a private pilot training under Part 61 before application for a commercial pilot certificate? (14 CFR 61.129)

Total Time: 250 hours of flight time as a pilot that consists of at least—

100 hours in powered aircraft, of which 50 hours must be in airplanes.

100 hours of pilot-in-command flight time, that includes at least:

a. 50 hours in airplanes; and

b. 50 hours in cross-country flying of which at least 10 hours must be in airplanes.

Dual: 20 hours of flight training on the Commercial Pilot areas of operation that includes at least:

a. 10 hours of instrument training of which at least 5 hours must be in a single-engine airplane;

b. 10 hours of training in a complex airplane or a turbine-powered airplane;

c. 1 cross-country of 2 hours in a single-engine airplane in day VFR conditions of a total straight-line distance greater than 100 NM from the departure point;

d. 1 cross-country of 2 hours in a single-engine airplane in night VFR conditions of a total straight-line distance greater than 100 NM from the departure point;

e. 3 hours of flight training in a single-engine airplane within the preceding 60 days prior to the practical test.

Continued

Solo: 10 hours of solo flight in a single-engine airplane on the Commercial Pilot areas of operation, including:

a. One cross-country flight of not less than 300 NM with landings at a minimum of 3 points, one of which is a straight-line distance of at least 250 NM; and

b. 5 hours in night VFR conditions with 10 takeoffs and landings (each landing must involve flight in the traffic pattern) at a controlled airport.

5. What privileges and limitations apply to commercial pilots? (14 CFR 61.133)

Commercial pilot general privileges are as follows: a person who holds a commercial pilot certificate may act as pilot-in-command of an aircraft carrying persons or property for compensation or hire, provided the person is qualified in accordance with 14 CFR Part 61, and for compensation or hire, provided the person is qualified in accordance with Part 61 and applicable parts of the regulations that apply to the operation.

As for limitations, a person who applies for a commercial pilot certificate with an airplane category or powered-lift category rating and does not hold an instrument rating in the same category and class will be issued a commercial pilot certificate that contains the limitation, "The carriage of passengers for hire in (airplanes) (powered-lifts) on cross-country flights in excess of 50 nautical miles or at night is prohibited." The limitation may be removed when the person satisfactorily accomplishes the requirements listed in §61.65 for an instrument rating in the same category and class of aircraft listed on the person's commercial pilot certificate.

G. Airline Transport Pilots

1. What are the general eligibility requirements for an airline transport pilot certificate? (14 CFR 61.153)

a. Be at least 23 years of age;

b. Be able to read, speak, write, and understand the English language;

c. Be of good moral character;

    d. Meet at least one of the following requirements:

        i. Hold at least a commercial pilot certificate and an instrument rating;

        ii. Meet the military experience requirements under §61.73 to qualify for a commercial pilot certificate, and an instrument rating if the person is a rated military pilot or former rated military pilot of an Armed Force of the United States; or

        iii. Hold either a foreign airline transport pilot or foreign commercial pilot license and an instrument rating if the person holds a pilot license issued by a contracting State to the Convention on International Civil Aviation;

    e. Meet the aeronautical experience requirements applicable to the aircraft category and class rating sought before applying for the practical test;

    f. Pass a knowledge test on the required aeronautical knowledge areas that apply to the aircraft category and class rating sought;

    g. Pass the practical test on the required areas of operation that apply to the aircraft category and class rating sought; and

    h. Comply with the sections of Part 61 that apply to the aircraft category and class rating sought.

2. What minimum aeronautical experience must be accumulated by a person applying for an Airline Transport Pilot certificate with an airplane category and class rating? (14 CFR 61.159)

Total Time: 1,500 hours that includes —

a. 500 hours of cross-country flying;

b. 100 hours of night flight time;

c. 75 hours of instrument time, actual or simulated; and

d. 250 hours in an airplane as a PIC, or as SIC performing the duties of PIC, or any combination thereof, that includes at least:

    i. 100 hours of cross-country flying; and

    ii. 25 hours of night flight time.

H. Flight Instructors

1. In what areas of aeronautical knowledge must an applicant for a flight instructor certificate have received instruction? (14 CFR 61.185)

The fundamentals of instructing, including:

a. The learning process;

b. Elements of effective teaching;

c. Student evaluation and testing;

d. Course development;

e. Lesson planning; and

f. Classroom training techniques.

Also, the aeronautical knowledge areas for a recreational, private, and commercial pilot certificate applicable to the aircraft category for which flight instructor privileges are sought; and the aeronautical knowledge areas for the instrument rating applicable to the category for which instrument flight instructor privileges are sought.

2. In what areas of operation must an applicant for a flight instructor certificate have received instruction? (14 CFR 61.187)

a. Fundamentals of instructing

b. Technical subject areas

c. Preflight preparation

d. Preflight lesson on a maneuver to be performed in flight

e. Preflight procedures

f. Airport and seaplane base operations

g. Takeoffs, landings, and go-arounds

h. Fundamentals of flight

i. Performance maneuvers

j. Ground reference maneuvers

k. Slow flight, stalls, and spins

l. Basic instrument maneuvers

m. Emergency operations

n. Postflight procedures

3. What are the required records a flight instructor must retain? (14 CFR 61.189)

a. A flight instructor must sign the logbook of each person to whom that instructor has given flight training or ground training.

b. A flight instructor must maintain a record in a logbook or a separate document that contains the name of each person whose logbook or student pilot certificate that instructor has endorsed for solo flight privileges, and the date of the endorsement; and the name of each person that instructor has endorsed for a knowledge test or practical test, and the record shall also indicate the kind of test, the date, and the results.

c. Each flight instructor must retain the required records for at least 3 years.

4. What are the various endorsements a flight instructor is authorized to give? (14 CFR 61.193)

A person who holds a flight instructor certificate is authorized within the limitations of that person's flight instructor certificate and ratings, and that person's pilot certificate and ratings, to give training and endorsements that are required for, and relate to:

a. A student pilot certificate;

b. A pilot certificate;

c. A flight instructor certificate;

d. A ground instructor certificate;

e. An aircraft rating;

f. An instrument rating;

g. A flight review, operating privilege, or recency of experience requirement;

h. A practical test; and

i. A knowledge test.

5. What limitations on endorsements apply to all flight instructors? (14 CFR 61.195)

A flight instructor may *not* endorse:

a. A student pilot's certificate or logbook for solo flight privileges, unless that flight instructor has given that student the required flight training and has determined that the student is prepared to conduct the flight safely under known circumstances, subject to any limitations that the instructor considers necessary for the safety of the flight;

b. A student pilot's certificate and logbook for a solo cross-country flight, unless that flight instructor has determined the student's flight preparation, planning, equipment, and proposed procedures are adequate for the proposed flight under the existing conditions and within any limitations that the instructor considered necessary for the safety of the flight;

c. A student pilot's certificate and logbook for solo flight in a Class B airspace area or at an airport within Class B airspace unless that flight instructor has given that student ground and flight training in that Class B airspace or at that airport; and determined that the student is proficient to operate the aircraft safely;

d. The logbook of a recreational pilot, unless that flight instructor has given that pilot the required ground and flight training; and determined that the recreational pilot is proficient to operate the aircraft safely;

e. The logbook of a pilot for a flight review, unless that instructor has conducted a review of that pilot in accordance with the requirements of §61.56(a);

f. The logbook of a pilot for an instrument proficiency check, unless that instructor has tested that pilot in accordance with the requirements of §61.57(d).

6. How many hours of flight training is a flight instructor limited to? (14 CFR 61.195)

In any 24-consecutive-hour period, a flight instructor may not conduct more than 8 hours of flight training.

7. What qualifications must a flight instructor possess before instruction may be given for the issuance of an instrument rating? (14 CFR 61.195)

This flight instructor must hold an instrument rating or a type rating not limited to VFR and must hold an instrument rating on his or her flight instructor certificate and pilot certificate appropriate to the category and class of aircraft in which instrument training is being provided.

8. What is the minimum pilot-in-command time requirement for a flight instructor with multi-engine privileges to give instruction to a student for a multi-engine rating? (14 CFR 61.195)

That flight instructor must have at least 5 flight hours of pilot-in-command time in the specific make and model of multi-engine airplane, helicopter, or powered-lift, as appropriate.

9. How can a flight instructor certificate be renewed? (14 CFR 61.197)

A person who holds a current flight instructor certificate may renew that certificate for an additional 24 calendar months if the holder:

a. Has passed a practical test for one of the ratings listed on the current flight instructor certificate, or an additional flight instructor rating; or

b. Has presented to an authorized FAA Flight Standards Inspector a record of training students showing that, during the preceding 24 calendar months, the flight instructor has endorsed at least five students for a practical test for a certificate or rating, and at least 80 percent of those students passed that test on the first attempt; or has a record showing that, within the preceding 24 calendar months, the flight instructor served as a company check pilot, chief flight instructor, company check airman, or flight instructor in a Part 121 or Part 135 operation, or in a position involving the regular evaluation of pilots; or has a

Continued

graduation certificate showing that, within the preceding 3 calendar months, the person has successfully completed an approved flight instructor refresher course consisting of ground or flight training, or a combination of both.

Additional Study Questions:

1. If requested, who may inspect your airman certificate, medical certificate, or logbook? (14 CFR 61.3) L

2. Must an airman who serves as safety pilot have a current medical certificate in their possession? (14 CFR 61.23, 91.109)

3. For what reasons may an examiner discontinue a practical test? (14 CFR 61.43)

4. Concerning instructional flights with both an authorized flight instructor and a certificated pilot onboard, which person is allowed to log pilot-in-command time? (14 CFR 61.51)

5. Under Part 61, must the CFI have his/her instrument rating (CFII) to teach the 3 hours of instrument training required for Private Pilot certificate? (14 CFR 61.109, 61.193)

6. Are appropriate ground instructor ratings necessary for a flight instructor to teach a ground school course? (14 CFR 61.193)

7. Can dual instruction be given to a pilot seeking an instrument rating from a flight instructor who does not possess an instrument flight instructor rating? (14 CFR 61.195) NO

8. After renewing your CFI certificate, you notice you also need a Biennial Flight Review. Does the CFI renewal satisfy the Biennial Flight Review requirement? (14 CFR 61.56) 1/2

Technical
Subject Areas

3

P + One Other

A. Aeromedical Factors

1. How can a student pilot obtain a medical certificate? (14 CFR 67.23)

By using Form 8500-8, "Application for Airman Medical Certificate or Airman Medical and Student Pilot Certificate." This form is filled out by the student and the Aviation Medical Examiner (AME). A list of AMEs is available at most Flight Standard District Offices.

2. Why is it important for students to obtain a medical certificate as soon as possible? (AIM 8-1-1)

Student pilots should visit an Aviation Medical Examiner as soon as possible in their flight training in order to avoid unnecessary training expenses should they not meet the medical standards.

3. How can an individual obtain a medical certificate in the event of a possible medical deficiency? (14 CFR 67.27)

Any person who is denied a medical certificate by an aviation medical examiner may, within 30 days after the date of denial, apply in writing and in duplicate to the Federal Air Surgeon for reconsideration of that denial. Aviation Medical Examiners (AME) can issue a medical certificate with certain limitations on flying activities due to medical conditions that exist.

4. State several medical conditions which might prevent the issuance of a medical certificate. (14 CFR Part 67)

a. Clinical diabetes

b. Coronary heart disease/heart attack

c. Epilepsy

d. Disturbance of consciousness

e. Alcoholism

f. Drug dependence

g. Psychosis

5. What should students know about flight operations conducted while suffering from a medical deficiency? (14 CFR 61.53)

No person may act as pilot-in-command, or in any other capacity as a required flight crewmember, while having a known medical deficiency that would make him/her unable to meet the requirements for his/her current medical certificate.

Additional Study Questions:

1. What are the symptoms, effects, and corrective actions for the following: hypoxia, hyperventilation, middle ear and sinus problems, carbon monoxide poisoning? (AC 67-2)

2. What are the causes, effects, and corrective actions for the following: spatial disorientation, motion sickness? (AC 67-2)

3. Explain the effects of alcohol and drugs and their relationship to safety and flying. (AC 67-2)

4. Explain the effect of nitrogen excesses during scuba dives and how this affects a pilot during flight. (AC 67-2)

5. Explain the effects of "fatigue" on the safety of flight and the corrective actions to prevent it. (AC 67-2)

B. Visual Scanning and Collision Avoidance

1. What is the relationship between a pilot's physical or mental condition and vision? (AIM 8-1-1)

The two are closely related. Even a minor illness suffered in day-to-day living can seriously degrade performance of many piloting tasks including the pilot's ability to effectively use his/her vision. Use the IMSAFE checklist: I'm physically and mentally safe to fly, not being impaired by: Illness, Medication, Stress, Alcohol, Fatigue, Emotion.

2. Name several factors that can degrade a pilot's vision. (AIM 8-1-6, 8-1-8)

a. *Visibility conditions* — smoke, haze, dust, rain, and flying towards the sun can greatly reduce the ability to detect targets.

b. *Windshield conditions* — dirty or bug-smeared windshields can greatly reduce the ability to see other aircraft.

c. *Bright illumination* — reflected off clouds, water, snow, and desert terrain that produces glare resulting in eye strain and the inability to see effectively.

d. *Dim illumination* — small print and colors on aeronautical charts and aircraft instruments become unreadable.

e. *Dark adaptation* — eyes must have at least 20 to 30 minutes to adjust to reduced light conditions.

3. Describe some of the various optical illusions a pilot can experience in flight. (AIM 8-1-4)

The "leans" — An abrupt correction of a banked attitude, which has been entered too slowly to stimulate the motion sensing system in the inner ear, can create the illusion of banking in the opposite direction. Examples include: Coriolis illusion, graveyard spin, graveyard spiral, somatogravic illusion, inversion illusion, elevator illusion, false horizon and autokinesis.

Runway width illusion — A narrower-than-usual runway can create the illusion that the aircraft is at a higher altitude than it actually is. Unrecognized, the pilot may fly a lower approach, with the risk of striking objects along the approach path or landing short. A wider-than-usual runway can have the opposite effect, with the risk of leveling out high and landing hard or overshooting the runway.

Runway and terrain slope illusion — An upsloping runway, upsloping terrain or both, can create the illusion that the aircraft is at a higher altitude than it actually is. Unrecognized, the pilot may fly a lower approach. A downsloping runway, downsloping approach terrain, or both, can have the opposite effect.

Featureless terrain illusion — An absence of ground features, as when landing over water, darkened areas, and terrain made featureless by snow, can create the illusion that the aircraft is at a higher altitude than it actually is. Unrecognized, the pilot may fly a lower approach.

4. Explain the "see & avoid" concept. (14 CFR 91.113)

When weather conditions permit, regardless of whether the flight is conducted under instrument flight rules or visual flight rules, vigilance shall be maintained by each person operating an aircraft so as to see and avoid other aircraft.

5. Explain the practice of time sharing your attention inside and outside the cockpit. (AIM 8-1-6)

Studies show that the time a pilot spends on visual tasks inside the cabin should represent no more than 1/4 to 1/3 of the scan time outside, or no more than 4 to 5 seconds on the instrument panel for every 16 seconds outside.

6. What is a good visual scanning technique? (AIM 8-1-6)

Effective scanning is accomplished with a series of short, regularly spaced eye movements that bring successive areas of the sky into the central visual field. Each movement should not exceed 10 degrees, and each area should be observed for at least 1 second to enable detection. Although horizontal back and forth eye movements seem preferred, each pilot should develop a scanning pattern that is most comfortable and then adhere to it to ensure optimum scanning.

Additional Study Questions:

1. Explain the relationship between aircraft speed differential and collision risk. (AC 90-48)

2. What situations involve the greatest collision risk? (AIM 8-1-8)

3. Explain the importance of knowing the aircraft's blind spots. (AIM 8-1-8)

4. Describe some of the situations which involve the greatest collision risk. (AIM 8-1-8)

5. What are proper clearing procedures? (AC 61-23)

C. Use of Distractions During Flight Training

1. What are several examples of flight situations where pilot distraction was a causative factor related to a stall/spin accident? (AC 61-67)

a. Engine failure.

b. Preoccupation inside or outside the cockpit while changing power, configuration, or trim.

c. Maneuvering to avoid other traffic.

d. Clearing hazardous obstacles during takeoff and climb.

2. Why is it important for instructors to incorporate realistic distractions during flight training? (AC 61-67)

Distractions should be included in the flight evaluation to determine that students possess the skills required to cope with distractions while maintaining aircraft control required for safe flight.

3. Give several examples of realistic distractions for specific flight situations you might utilize during flight training. (AC 61-67)

a. Simulate frequency changes and communications.

b. Read outside air temperature gauge.

c. Remove object from glove compartment or map case.

d. Identify terrain features or objects on the ground.

Additional Study Questions:

1. What is the difference between the proper use of distractions and harassment? (AC 61-67)

2. Describe the relationship between division of attention and flight instructor use of distractions. (AC 61-67)

D. Principles of Flight

1. What are the four dynamic forces that act on airplane during all maneuvers? (AC 61-23)

The airplane in straight-and-level unaccelerated flight is acted on by four forces:

Lift—the upward acting force

Gravity—or weight, the downward acting force

Thrust—the forward acting force

Drag—the backward acting force

2. Explain the following terms as they relate to the production of lift. (AC 61-23)

Bernoulli's principle—states in part that the internal pressure of a fluid (liquid or gas) decreases at points where the speed of the fluid increases. In other words, high speed flow (above the wing) is associated with low pressure, and low speed flow (below the wing) with high pressure.

Newton's Third Law—since for every action there is an opposite and equal reaction, an additional upward force is generated as the lower surface of the wing deflects air downward.

3. Define the following terms: (AC 61-23)

Airfoil—any surface designed to obtain reaction, such as lift, from the air through which it moves.

Angle of attack—the angle between the chord line of the wing and the direction of the relative wind.

Angle of incidence—the angle formed by the chord line of the wing and the longitudinal axis of the airplane.

Camber—the curvature of the airfoil from the leading edge to the trailing edge.

Chord line—an imaginary straight line drawn from the leading edge to the trailing edge of a cross section of an airfoil.

Wing planform—the shape or form of a wing as viewed from above. It may be long and tapered, short and rectangular, or various other shapes.

4. Explain the forces acting on an airplane when operating at an airspeed slower than cruise. (AC 61-23)

While maintaining straight-and-level flight at constant airspeeds slower than cruise, the opposing forces must still be equal in magnitude. However, some of these forces are separated into components. In this flight condition the actual throughst no longer acts parallel and opposite to the flightpath and drag. Actual throughst is inclined upward. Actual throughst now has two components: one acting perpendicular to the flightpath in the direction of lift, the other acting along the flightpath. Because the actual throughst is inclined, its magnitude must be greater than drag if its component of throughst along the flightpath is equal to drag. To summarize, in straight-and-level flight at slow airspeeds, the actual throughst is greater than drag and wing lift is less than at cruise airspeed.

5. What causes an airplane to turn? (AC 61-23)

First, keep in mind that the rudder does not turn the airplane in flight. The airplane must be banked because the same force that sustains the airplane in flight is used to make the airplane turn. The airplane is banked and back elevator pressure is applied. This changes the direction of lift and increases the angle of attack on the wings, which increases the lift. The increased lift pulls the airplane around the turn. In a turn, the force of lift can be resolved into two components: vertical and horizontal. During the turn entry, the vertical component of lift still opposes gravity, and the horizontal component must overcome centrifugal force. Consequently, the total lift must be sufficient to counteract both of these forces. The airplane is pulled around the turn, not sideways, because the tail section acts as a weathervane which continually keeps the airplane streamlined with the curved flightpath.

6. Define stability, controllability, and maneuverability. (AC 61-23)

Stability is the aircraft's ability to return to normal flight after being disturbed by such events as turbulence. There are two basic types: static and dynamic.

Controllability is the aircraft's ability to respond to the pilot's control inputs.

Maneuverability is the aircraft's structural ability to be handled easily.

7. Explain the effect of wing downwash on longitudinal stability. (AC 61-23)

In level flight, there will be a downwash of air from the wings. This downwash strikes the top of the horizontal stabilizer and produces a downward pressure, which at a certain speed will be just enough to balance the inherent nose-heaviness of an airplane. The faster the airplane is flying, the greater the downwash and the greater the downward force on the horizontal stabilizer. If the airplane's speed decreases, the speed of the airflow over the wing is decreased. As a result of this decreased airflow over the wing, the downwash is reduced, causing a lesser downward force on the horizontal stabilizer. In turn, the characteristic nose-heaviness is accentuated, causing the airplane's nose to pitch down more. This places the airplane in a nose-low attitude, lessening the wings' angle of attack and drag, and allowing the airspeed to increase. As the airplane continues in the nose-low attitude and its speed increases, the downward force on the horizontal stabilizer is once again increased. Consequently the tail is again pushed downward and the nose rises into a climbing attitude.

8. What are the four factors that contribute to torque effect? (AC 61-23)

a. *Torque reaction of the engine and propeller:* For every action there is an equal and opposite reaction. The rotation of the propeller (from the cockpit) to the right, tends to roll or bank the airplane to the left.

b. *Gyroscopic effect of the propeller:* Gyroscopic precession applies here, the resultant action or deflection of a spinning object when a force is applied to the outer rim of its rotational mass. If the axis of a propeller is tilted, the resulting force will be exerted 90 degrees ahead in the direction of rotation and in the same direction as the applied force. Most noticeable on takeoffs in taildraggers.

Remember: Nose yaws to the right when nose is moving upwards (tail down). Nose yaws to the left when nose is moving downwards (tail up).

c. *Corkscrewing effect of the propeller slipstream:* High speed rotation of an airplane propeller results in a corkscrewing rotation to the slipstream as it moves rearward. Most noticeable at high propeller speeds and low forward speeds, (as in a takeoff), the slipstream strikes the vertical tail surface on the left side pushing the tail to the right and yawing the airplane to the left.

d. *Asymmetrical loading of the propeller (P-Factor):* When an airplane is flying with a high angle of attack, the bite of the downward-moving propeller blade is greater than the bite of the upward moving blade. This is due to the downward-moving blade meeting the oncoming relative wind at a greater angle of attack than the upward-moving blade. Consequently there is greater throughst on the downward moving blade on the right side, and this forces the airplane to yaw to the left.

9. Explain the effects of speed on load factor. (AC 61-23)

The amount of excess load that can be imposed on the wing depends on how fast the airplane is flying. At slower speeds, the maximum allowable lifting force of the wing is only slightly greater than the amount necessary to support the weight of the airplane. Consequently, the load factor should not become excessive even if the controls are moved abruptly or the airplane encounters severe gusts. The reason for this is the airplane will stall before the load can become excessive. However, at high speeds, the lifting capacity of the wing is so great that a sudden movement of the elevator controls or a strong gust may increase the load factor beyond safe limits.

Additional Study Questions:

1. **What are several factors which will affect both lift and drag?** (AC 61-23)

2. **Explain the distribution of forces in the following phases of flight: straight-and-level, climbs, descents, turns.** (AC 61-23)

3. **Explain "ground effect."** (H-8083-3)

4. **Explain the relationship of center of gravity and center of pressure to longitudinal stability.** (AC 61-23)

5. **When lowering the flaps, why do some aircraft encounter a pitch change?** (AC 61-23)

6. **What are several methods an aircraft manufacturer may employ to correct for turning tendency or torque effect in flight?** (AC 61-23)

7. **What is a "Vg" diagram?** (AC 61-23)

8. **What are wingtip vortices and what flight precautions should be practiced?** (AC 61-23)

E. Elevators, Ailerons, and Rudder

1. **What are elevators and what is their function?** (H-8083-3)

The elevators are attached to the horizontal portion of the empennage—the horizontal stabilizer. The elevators provide control of the pitch attitude about the airplane's lateral axis.

2. **What are ailerons and what is their function?** (H-8083-3)

The movable portions of each wing are the ailerons. They are located on the trailing edge of each wing near the outer tips. When deflected up or down they, in effect, change the wing's camber (curvature) and its angle of attack, and therefore change the wing's lift/drag characteristics. Their primary use is to bank (roll) the airplane around its longitudinal axis.

3. What is a rudder and what is its function? (H-8083-3)

The rudder is attached to the fixed vertical portion of the empennage—the vertical fin or vertical stabilizer. It is used to control direction (left or right) of yaw about the airplane's vertical axis.

Additional Study Questions:

1. Describe the system of control for the elevators, ailerons and rudder on your airplane. (AFM)

2. What are differential-type ailerons? (H-8083-3)

3. What is "adverse yaw"? (H-8083-3)

F. Trim Devices

1. What are movable trim devices and what is their function? (H-8083-3)

A trim tab is a small auxiliary control surface hinged at the trailing edge of a primary flight control surface (elevator, aileron, rudder). They are commonly used to relieve the pilot of maintaining continuous pressure on the primary controls when correcting for an unbalanced flight condition resulting from changes in aerodynamic forces or weight. A trim tab acts on the primary flight control surface, which in turn acts upon the entire airplane. A trim tab is a part of the control surface but may be moved up or down independently of the surface itself.

2. **Describe the direction of movement of the elevator trim tab and elevator when the pilot is applying "up" or "down" elevator trim.** (H-8083-3)

An upward deflection of the trim tab will force the elevator downward with the same result as moving the elevator downward with the elevator control, and conversely a downward deflection of the trim tab will force the elevator upward. The direction the trim tab is deflected will always cause the entire elevator to be deflected in the opposite direction.

Additional Study Questions:

1. **Describe the location, attachments, and system of control for the trim system on your airplane.** (AFM)

2. **Explain the proper technique for use of trim on your airplane.** (AFM)

3. **What are "fixed" trim tabs and where are they normally found on aircraft?** (AFM)

G. Wing Flaps

1. **What are flaps and what is their function?** (AC 61-23)

The wing flaps are movable panels on the inboard trailing edges of the wings. They are hinged so that they may be extended downward into the flow of air beneath the wings to increase both lift and drag. Their purpose is to permit a slower airspeed and a steeper angle of descent during a landing approach. In some cases, they may also be used to shorten the takeoff distance.

2. **Name four basic types of flaps.** (AC 61-23)

Plain flap—Most common flap system; when extended it increases the camber of the wing which results in an increase in both lift and drag.

Split flap—Similar to the plain flap; the main difference is the split flap produces more drag which enables a steeper approach without an increase in airspeed.

Slotted flap—A slotted flap will produce proportionally more lift than drag. Its design allows high pressure air below the wing to be directed through a slot to flow over the upper surface of the flap, delaying the airflow separation at higher angles of attack. This design lowers the stall speed significantly.

Fowler flap—Very efficient design. Moves backward on first part of extension increasing lift with little drag; also utilizes a slotted design resulting in lower stall speeds.

3. When lowering the flaps, why do some aircraft encounter a pitch change? (AC 61-23)

Use of flaps alters the lift distribution and consequently causes the center of pressure to move aft. Because of this movement, a nose-down pitching moment may be experienced. Low-wing aircraft are more subject to this characteristic than high-wing airplanes. With flap extension on a high-wing airplane, the center of pressure also moves aft but the airflow from the wing is directed downward which pushes down on the horizontal stabilizer. This results in a counteracting nose-up pitching moment.

Additional Study Questions:

1. Describe the wing flap system for your airplane. (AFM)

2. Explain the proper technique for use of flaps in your airplane. (AFM)

3. Explain how the extension of flaps will affect stall speed. (H-8083-3)

4. Explain the procedure for dealing with a split flap situation. (AFM)

H. Airplane Weight and Balance

1. Define the following weight and balance terms:

Arm; Basic Operating Weight; Center of Gravity; Center of Gravity Limits; Center of Gravity Range; Datum; Empty Weight; Fuel Load; Moment; Moment Index; Station; Useful Load. (H-8083-1)

Arm—The horizontal distance in inches from the reference datum line to the center of gravity of an item.

Basic Operating Weight—The weight of the aircraft, including the crew, ready for flight but without payload and fuel. This term is only applicable to transport aircraft.

Center of Gravity—The point about which an aircraft would balance if it were possible to suspend it at that point, expressed in inches from datum.

Center of Gravity Limits—The specified forward and aft or lateral points beyond which the CG must not be located during takeoff, flight or landing.

Center of Gravity Range—The distance between the forward and aft CG limits indicated on pertinent aircraft specifications.

Datum—An imaginary vertical plane or line from which all measurements of arm are taken. It is established by the manufacturer.

Empty Weight—The airframe, engines, and all items of operating equipment that have fixed locations and are permanently installed in the aircraft. It includes optional and special equipment, fixed ballast, hydraulic fluid, unusable fuel, and undrainable oil.

Fuel Load—The expendable part of the load of the aircraft. It includes only usable fuel, not fuel required to fill the lines or that which remains trapped in the tank sumps.

Moment—The product of the weight of an item multiplied by its arm. Moments are expressed in pound-inches.

Moment Index—A moment divided by a constant such as 100, 1,000, or 10,000. The purpose of using a moment index is to simplify weight and balance computations of large aircraft where heavy items and long arms result in large, unmanageable numbers.

Station—A location in the aircraft which is identified by a number designating its distance in inches from the datum. The datum is, therefore, identified as station zero. The station and arm are usually identical. An item located at station +50 would have an arm of 50 inches.

Useful Load—The weight of the pilot, copilot, passengers, baggage, usable fuel and drainable oil. It is the empty weight subtracted from the maximum allowable takeoff weight. The term applies to general aviation aircraft only.

2. What performance characteristics will be adversely affected when an aircraft has been overloaded? (AC 61-23)

a. Higher takeoff speed

b. Longer takeoff run

c. Reduced rate and angle of climb

d. Lower maximum altitude

e. Shorter range

f. Reduced cruising speed

g. Reduced maneuverability

h. Higher stalling speed

i. Higher landing speed

j. Longer landing roll

k. Excessive weight on the nosewheel

3. What effect does a forward center of gravity have on an aircraft's flight characteristics? (AC 61-23)

Higher stall speed—Stalling angle of attack reached at a higher speed due to increased wing loading.

Slower cruise speed—Increased drag, greater angle of attack required to maintain altitude.

More stable—When angle of attack is increased, the airplane tends to reduce angle of attack; longitudinal stability is improved.

Greater back elevator pressure required—Longer takeoff roll, higher approach speeds and problems with the landing flare.

4. **What effect does an aft center of gravity have on an aircraft's flight characteristics?** (AC 61-23)

   Lower stall speed—Less wing loading.

   Higher cruise speed—Reduced drag, smaller angle of attack required to maintain altitude.

   Less stable—Stall and spin recovery more difficult; when angle of attack is increased it tends to result in additional increased angle of attack.

5. **What basic equation is used in all weight and balance problems to find the center of gravity location of an airplane and/or its components?** (AC 61-23)

   Weight x Arm = Moment

   By rearrangement of this equation to the forms,

   Weight = Moment ÷ Arm

   Arm = Moment ÷ Weight

   CG = Moment ÷ Weight

   With any two known values, the third value can be found.

Additional Study Questions:

1. **If the weight of an aircraft is within takeoff limits but the CG limit has been exceeded, what actions can the pilot take to correct the situation?** (AC 61-23)

2. **When a shift in weight is required, what standardized and simple calculations can be made to determine the new CG?** (AC 61-23)

3. **If the weight of an aircraft changes due to the addition or removal of cargo or passengers before flight, what formula may be used to calculate the new CG?** (AC 61-23)

4. **What simple and fundamental weight check can be made before all flights?** (AC 61-23)

I. Navigation and Flight Planning

1. What are three common ways to navigate? (AC 61-23)

To navigate successfully, a pilot must know their approximate position at all times or be able to determine it whenever they wish. Position may be determined by:

a. Pilotage (reference to visible landmarks);

b. Dead reckoning (computing direction and distance from a known position); or

c. Radio navigation (use of radio aids).

2. Define the following terms: (AC 61-23)

Isogonic line—connect points of equal magnetic variation; found on most aeronautical charts and show amount and direction of variation.

Magnetic variation—the error induced by the difference in location of true north and magnetic north. Expressed in east or west variation.

Magnetic deviation—the amount of deflection of the compass needle caused by magnetic influences within the airplane (electrical circuits, radios, lights, tools, engine magnetized metal parts, etc.)

Lines of latitude—circles parallel to the equator (lines running east and west); used to measure distance in degrees north or south of the equator.

Line of longitude—lines drawn from the north pole to the south pole and are at right angles to the equator. The Prime Meridian is used as the zero reference line from which measurement is made in degrees east or west to 180 degrees.

Course—the intended path of an aircraft over the earth; or the direction of a line drawn on a chart representing the intended aircraft path.

Heading—direction the nose of the airplane points during flight.

Track—the actual path made over the ground in flight.

Drift angle—the angle between heading and track.

Wind correction angle—correction applied to the course to establish a heading so track will coincide with course.

3. What is the basic formula for determining a compass heading? (AC 61-23)

TC +/- WCA = TH +/- VAR = MH +/- DEV = CH

4. Explain, as you would to a student, the procedure for planning a VFR cross-country.

a. Get a preflight briefing consisting of the latest or most current weather, airport, and enroute NAVAID information.

b. Draw course lines and mark checkpoints on the chart.

c. Enter checkpoints on the log.

d. Enter NAVAIDs on the log.

e. Enter VOR courses on the log.

f. Enter altitude on the log.

g. Enter the wind (direction/velocity) and temperature on the log.

h. Measure the true course on the chart and enter it on the log.

i. Compute the true airspeed and enter it on the log.

j. Compute the WCA and groundspeed and enter them on the log.

k. Determine variation from chart and enter it on the log.

l. Determine deviation from compass correction card and enter it on the log.

m. Enter compass heading on the log.

n. Measure distances on the chart and enter them on the log.

o. Figure ETE and ETA and enter them on the log.

p. Calculate fuel burn and usage; enter them on the log.

q. Compute weight and balance.

r. Compute takeoff and landing performance.

s. Complete a Flight Plan form.

t. File the Flight Plan with FSS.

5. Explain the procedure for diverting to an alternate. (AC 61-23)

a. Mark your present position on the chart; write the current time next to your mark.

b. Consider the relative distance to all suitable alternates; select the one most appropriate for emergency.

c. Determine the magnetic course to the alternate and divert immediately.

d. Wind correction, actual distance and estimated time/fuel can then be computed while en route to alternate.

Note: Use the radial of a nearby VOR or airway that most closely parallels the course to the alternate. Distances can be determined by placing a finger at the appropriate place on a straight edge of a piece of paper and then measuring the approximate distance on the mileage scale at the bottom of the chart.

Additional Study Questions:

1. **Explain the various terms and symbols on a sectional chart.**

2. **Explain the different types of airspace on a sectional chart.**

3. **What are several types of radio aids to use for navigation?** (AC 61-23)

4. **What is a "VOR" and explain, as you would to a student, how to use it for navigation.** (AC 61-23)

5. **Explain the purpose of and the procedure used in filing a flight plan.** (AC 61-23)

6. **What procedures will you teach your student to use, in the event they should become lost on a cross-country flight?**

7. **Explain the procedure for calculating fuel consumption.** (AC 61-23)

8. **Explain the importance of preparing and properly using a flight log.** (AC 61-23)

J. Night Operations

1. What is the function of the rods and cones, and where are they located in the eye? (AC 61-23)

The rods are used for peripheral vision and are located in a ring around the cones. Rods are highly sensitive and are activated by a minimum amount of light. Thus they are better suited for night vision. The cones are used to detect color, detail and far-away objects and are located in the center of the retina at the back of the eye. They are less sensitive to light, require higher levels of intensity to become active, and are most useful in the daylight hours. The cones will take approximately 10–15 minutes to adjust to darkness. The rods will take approximately 30 minutes to adjust to darkness.

2. What can the pilot do to improve the effectiveness of vision at night? (AIM 8-1-6)

a. If possible, allow 30 minutes for the eyes to adjust to the darkness.

b. Adjust the cockpit lights for optimum night vision and avoid bright levels of light.

c. Never look directly at an object; try to look off center since this allows better visual perception.

d. Be aware that depth perception is inhibited due to lack of visual cues, thus attention to airspeed, altitude, sink rates and attitude indicators must be maintained.

e. Maintain a safe altitude until airport lighting and the airport itself are identifiable and visible. Many pilots have mistaken lighted highways for airports.

f. Avoid taking medication, drugs and alcohol prior to or during aircraft flights.

3. During preflight, what things should be done to adequately prepare for a night flight? (H-8083-3)

a. Study all weather reports and forecasts. Pay particular attention to temperature/dew point spreads to detect the possibility of fog formation.

b. Calculate wind directions and speeds along the proposed route of flight to ensure accurate drift calculations, as night visual perception of drift is generally inaccurate.

c. Obtain applicable aeronautical charts for both the proposed route as well as adjacent charts and clearly mark lighted checkpoints.

d. Review all radio navigational aids for correct frequencies and availability.

e. Check all personal equipment such as flashlights and portable transceivers for proper operation.

f. Do a thorough preflight inspection of the aircraft.

g. Check all aircraft position lights, as well as the landing light and rotating beacon, for proper operation.

h. Check ground areas for obstructions that may not be readily visible from the cockpit.

4. What are some guidelines to follow during the starting, taxiing, and run-up phases of a night flight? (H-8083-3)

a. The pilot should exercise extra caution on "clearing" the propeller arc area. The use of lights prior to engine startup can also alert persons in the area to the presence of the active aircraft.

b. During taxiing, avoid unnecessary use of electrical equipment which would put an abnormal load on the electrical system, such as the landing light.

c. Taxi slowly and follow any taxi lines.

5. What are some guidelines to follow during takeoff and departure phases of a night flight? (H-8083-3)

During takeoff the pilot should, on initial takeoff roll, use the distant runway edge lights as well as the landing light area to keep the aircraft straight and parallel with the runway on initial takeoff roll; and upon liftoff, keep a positive climb by referencing the attitude indicator along with positive rate of climb on the vertical speed indicator.

During climbout the pilot should not initiate any turns until reaching a safe maneuvering altitude, and should turn the landing light off as soon as possible.

6. What procedures should be followed during the approach and landing phase of a night flight? (H-8083-3)

a. The pilot should identify the airport and associated airport lighting and runway lighting.

b. The aircraft should be flown towards the airport beacon until the runway lights are identified.

c. A powered approach should be used, as visual perception during a descent at night can be difficult.

d. The landing light should be switched on upon entering the airport traffic area.

e. The pilot should avoid the use of excessive speed on approach and landing.

Additional Study Questions:

1. When approaching a well-lighted runway surrounded by a dark area with little or no features, what illusion should a pilot be alert for? (AIM 8-1-5)

2. If an engine failure occurs at night, what procedures should be followed? (H-8083-3)

3. Explain the procedure for conducting an approach and landing at night with no landing light. (H-8083-3)

4. **What equipment is required on an aircraft for night flights?** (14 CFR 91.205)

5. **Explain the function and color of the following airport lighting: runway edge lights, runway end identifier lights (REIL), taxiway edge lights, taxiway centerline lights.** (AIM 2-1-1 through 2-1-9)

6. **What additional equipment should the pilot have in the cockpit for night flight operations?** (H-8083-3)

K. High Altitude Operations

1. **What are the regulations concerning use of supplemental oxygen onboard an aircraft?** (14 CFR 91.211)

   No person may operate a civil aircraft of U.S. registry:

   a. At cabin pressure altitudes above 12,500 feet MSL up to and including 14,000 feet MSL, unless, for that part of the flight at those altitudes that is more than 30 minutes, the required minimum flight crew is provided with and uses supplemental oxygen.

   b. At cabin pressure altitudes above 14,000 feet MSL, unless the required flight crew is provided with and uses supplemental oxygen for the entire flight time at those altitudes.

   c. At cabin pressure altitudes above 15,000 feet MSL, unless each occupant is provided with supplemental oxygen.

2. **What is a "pressurized" aircraft?** (H-8083-3)

   In a "pressurized" aircraft, the cabin, flight compartment, and baggage compartments are incorporated into a sealed unit which is capable of containing air under a pressure higher than outside atmospheric pressure. Pressurized air is pumped into this sealed fuselage by cabin superchargers which deliver a relatively constant volume of air at all altitudes up to a designed maximum. Air is released from the fuselage by a device called an outflow valve. Since the superchargers provide a constant inflow of air to the pressurized area, the outflow valve, by regulating the air exit, is the major controlling element in the pressurization system.

3. What are the two types of oxygen breathing systems normally in use? (AC 65-15)

The type of oxygen system used is determined by the regulator; the two types are continuous-flow and pressure-demand. Most general aviation aircraft use the continuous-flow type.

4. Can any kind of oxygen be used for aviator's breathing oxygen? (AC 65-15)

No, oxygen used for medical purposes or welding normally should not be used because it may contain too much water. The excess water could condense and freeze in the oxygen lines when flying at high altitudes. Specifications for "aviator's breathing oxygen" are 99.5% pure oxygen with not more than two milliliters of water per liter of oxygen.

5. What are the two types of decompression? (H-8083-3)

Explosive decompression—Cabin pressure decreases faster than the lungs can decompress. Most authorities consider any kind of decompression which occurs in less than 1/2 second as explosive and potentially dangerous. This type of decompression could only be caused by structural damage, material failure, or by a door "popping" open.

Rapid decompression—A change in cabin pressure where the lungs decompress faster than the cabin. There is no likelihood of lung damage in this case. This type could be caused by a failure or malfunction in the pressurization system itself, or through slow leaks in the pressurized area.

Additional Study Questions:

1. What are the regulations pertaining to the use of supplemental oxygen on board a "pressurized" aircraft? (14 CFR 91.211)

2. Describe a typical cabin pressure control system. (H-8083-3)

3. **What are the components of a cabin pressure control system?** (H-8083-3)

4. **How does a continuous-flow oxygen system operate?** (AC 65-15)

5. **What are the dangers of decompression?** (H-8083-3)

L. Federal Aviation Regulations

1. **What is the purpose and general content of the following parts of the Federal Aviation Regulations?**

   14 CFR Part 61   Certification: Pilots, Flight Instructors, and
                     Ground Instructors

   14 CFR Part 91   General Operating and Flight Rules

   14 CFR Part 141   Pilot Schools

2. **What is the purpose and general content of NTSB Part 830?**

   Notification and Reporting of Aircraft Accidents or Incidents and Overdue Aircraft and Preservation of Aircraft Wreckage, Mail, Cargo and Records

Additional Study Questions:

1. **How can a pilot obtain a current copy of the Federal Aviation Regulations?**

2. **How often are the regulations revised and how are pilots notified?**

3. **What is the purpose and general content of the following parts of the Federal Aviation Regulations: 14 CFR Parts 1, 43, 67, and 71.**

M. Use of Minimum Equipment Lists

1. What are "Minimum Equipment Lists"? (14 CFR 91.213)

A Minimum Equipment List (MEL) can be published when compliance with minimum equipment requirements is not necessary in the interest of safety under all conditions. Deviation from the equipment requirements of the regulations is maintained by alternate means. Experience has shown that with the various levels of redundancy designed into aircraft, operation of every system or component installed may not be necessary when the remaining operative equipment can provide an acceptable level of safety. The FAA-approved MEL includes only those items of equipment the Administrator finds may be inoperative and yet maintain an acceptable level of safety by appropriate conditions and limitations. It takes into consideration the operator's particular aircraft equipment configuration and operational conditions.

2. What limitations apply to aircraft operations conducted using MELs? (14 CFR 91.213)

Once approved and authorized, the individual operator's MEL permits operation of the aircraft with inoperative equipment. The MEL permits operation with inoperative equipment for the minimum period of time necessary until repair can be accomplished. It is important the repair be done at the earliest opportunity in order to return the aircraft to its design level of safety and reliability. The MEL establishes limitations on the duration of operation with inoperative equipment.

3. What responsibilities should a pilot be familiar with concerning operation of an aircraft utilizing a Minimum Equipment List? (14 CFR 91.213)

No person may take off in an aircraft with inoperative instruments or equipment installed unless

a. An approved Minimum Equipment List exists for that aircraft;

b. A letter of authorization from the FAA is carried within the aircraft authorizing use of a MEL;

c. The aircraft records available to the pilot must include an entry describing the inoperable instruments and equipment allowed by the MEL.

4. What length of time can an aircraft be flown with inoperative equipment onboard? (AC 91-67)

An operator may defer maintenance on inoperative equipment that has been deactivated or removed and placarded inoperative. When the aircraft is due for inspection in accordance with the regulations, the operator should have all inoperative items repaired or replaced. If an owner does not want specific inoperative equipment repaired, then the maintenance person must check each item to see if it conforms to the requirements of 14 CFR §91.213. The maintenance person must ensure that each item of inoperative equipment that is to remain inoperative is placarded appropriately.

5. What are "Special Flight Permits" and when are they necessary? (14 CFR 91.213, 21.197)

A "Special Flight Permit" may be issued for an aircraft that may not currently meet applicable airworthiness requirements but is capable of safe flight. These permits are typically issued for the following purposes:

a. Flying an aircraft to a base where repairs, alterations or maintenance are to be performed, or to a point of storage.

b. Delivering or exporting an aircraft.

c. Flight testing new production aircraft.

d. Evacuating aircraft from areas of impending danger.

e. Conducting customer demonstration flights in new production aircraft that have satisfactorily completed production flight tests.

Additional Study Questions:

1. **Are MELs available for all aircraft?** (14 CFR 91.213)

2. **If an aircraft is not being operated under a MEL, how can you determine which instruments and equipment onboard can be inoperative and the aircraft still be legal for flight?** (14 CFR 91.213)

N. Publications

1. What is the "Aeronautical Information Manual"? (AIM)

The *Aeronautical Information Manual* (AIM), produced by the FAA, is designed to provide the aviation community with basic flight information and ATC procedures for use in the National Airspace System of the United States. The manual also contains items of interest to pilots concerning health and medical facts, factors affecting flight safety, a pilot/controller glossary of terms used in the Air Traffic Control System, and information on safety, accident, and hazard reporting requirements.

2. What are Advisory Circulars? (AC 61-23)

Advisory Circulars are issued by the FAA and are used to distribute non-regulatory information to the public. The Government Printing Office produces an Advisory Circular checklist which provides a list of all available Advisory Circulars and the necessary forms for ordering. Advisory Circulars are issued in a numbered subject system corresponding to the subject areas of the Federal Aviation Regulations.

Example:

AC 00   General
AC 10   Procedural Rules
AC 20   Aircraft
AC 60   Airmen
AC 70   Airspace
AC 90   Air Traffic and General Operating Rules
AC 120   Air Carriers

3. **Explain the purpose and general content of the Practical Test Standards.**

   Practical Test Standards are published by the FAA to establish the standards for the various practical tests administered to obtain a pilot certificate or rating. FAA inspectors and designated pilot examiners conduct practical tests in compliance with these standards. Flight instructors and applicants should find these standards helpful in practical test preparation. The Practical Test Standards are available for purchase from the Government Printing Office as well as many commercial vendors who reprint them.

4. **Explain the availability, purpose, and general content of the Airport/Facility Directory.** (AC 61-23)

   The Airport/Facility Directory, produced by the National Ocean Service every 8 weeks, is a directory of all airports, seaplane bases and heliports open to the public. Also, communications information and other navigational data is made available. The Airport/Facility Directory is available for purchase (subscription) from the National Ocean Service and is also sold separately by many commercial vendors. The A/FD contains information such as abbreviations, airports facilities, notices, FAA and National Weather Service telephone numbers, Air Route Traffic Control Centers, Flight Service Station Communication Frequencies, FSDO addresses/telephone numbers, etc.

Additional Study Questions:

1. **What Advisory Circular should be referenced for the appropriate endorsements when recommending an applicant for a practical test?**

2. **What are NOTAMs and where can they be obtained?** (AC 61-23)

3. **Explain the differences between FAA Approved Airplane Flight Manuals, Pilot Operating Handbooks, and Owner's Manuals.**

O. National Airspace System

1. What is Class A airspace? (AIM 3-2-2)

Generally, that airspace from 18,000 feet MSL up to and including FL600, including that airspace overlying the waters within 12 nautical miles of the coast of the 48 contiguous states and Alaska; and designated international airspace beyond 12 nautical miles of the coast of the 48 contiguous states and Alaska within areas of domestic radio navigational signal or ATC radar coverage, and within which domestic procedures are applied.

2. What is Class B airspace? (AIM 3-2-3)

Generally, that airspace from the surface to 10,000 feet MSL surrounding the nation's busiest airports in terms of IFR operations or passenger enplanements. The configuration of each Class B airspace area is individually tailored and consists of a surface area and two or more layers (some Class B airspace areas resemble upside down wedding cakes), and is designed to contain all published instrument procedures once an aircraft enters the airspace.

3. What is Class C airspace? (AIM 3-2-4)

Generally, that airspace from the surface to 4,000 feet above the airport elevation (charted in MSL) surrounding those airports that have an operational control tower, are serviced by a radar approach control, and that have a certain number of IFR operations or passenger enplanements.

4. What is Class D airspace? (AIM 3-2-5)

Generally, that airspace from the surface to 2,500 feet above the airport elevation (charted in MSL) surrounding those airports that have an operational control tower. The configuration of each Class D airspace area is individually tailored and when instrument procedures are published, the airspace will normally be designed to contain those procedures.

5. What is Class E (controlled) airspace? (AIM 3-2-6)

Generally, if the airspace is not Class A, Class B, Class C, or Class D, and it is controlled airspace, it is Class E airspace. Except for 18,000 feet MSL, Class E airspace has no defined vertical limit but rather it extends upward from either the surface or a designated altitude to the overlying or adjacent controlled airspace.

6. What is Class G (uncontrolled) airspace? (AIM 3-3-1)

Class G, or uncontrolled, airspace is that portion of the airspace that has not been designated as Class A, B, C, D, and E airspace. Class G airspace begins at the surface and continues up to the overlying controlled (Class E) airspace, not to exceed 14,500 feet MSL.

7. Define the following types of special use airspace. (AIM 3-4-1 through 3-4-8)

Prohibited Area—For security or other reasons, aircraft flight is prohibited.

Restricted Area—Contains unusual, often invisible hazards to aircraft. Flights must have permission from the controlling agency if VFR. IFR flights will be cleared through or vectored around it.

Military Operations Area—Designed to separate military training from IFR traffic. Permission is not required but VFR flights should exercise caution. IFR flights will be cleared through or vectored around it.

Warning Area—Same hazards as a restricted area, it is established beyond the 3-mile limit of International airspace. Permission is not required but a flight plan is advised.

Alert Area—Airspace containing a high volume of pilot training or unusual aerial activity. No permission is required but VFR flights should exercise caution. IFR flights will be cleared through or vectored around it.

Controlled Firing Areas—CFAs contain activities which, if not conducted in a controlled environment, could be hazardous to non-participating aircraft. Activities are suspended immediately when

Continued

spotter aircraft, radar or ground lookout positions indicate an aircraft might be approaching the area. CFAs are not charted.

National Security Areas—Airspace of defined vertical and lateral dimensions established at locations where there is a requirement for increased security and safety of ground facilities. Pilots are requested to voluntarily avoid flying through the depicted NSA. When it is necessary to provide a greater level of security and safety, flight in NSAs may be temporarily prohibited by regulation under the provisions of 14 CFR § 99.7.

Additional Study Questions:

1. **Explain the general dimensions for each class of airspace.** (AIM 3-2-1 through 3-2-6)

2. **Give examples of "Other" airspace areas.** (AIM 3-5-1 through 3-5-7)

3. **What are the cloud clearance and visibility requirements to operate under VFR in controlled and uncontrolled airspace?** (14 CFR 91.155)

4. **Explain the requirements to operate under "Special VFR."** (14 CFR 91.157)

P. Logbook Entries and Certificate Endorsements

1. **As a CFI, what are the endorsements you will use for student pilots?** (AC 61-65)

   a. Pre-solo aeronautical knowledge

   b. Pre-solo flight training

   c. Pre-solo flight training at night

   d. Solo flight (each additional 90-day period)

   e. Solo takeoffs and landings at another airport within 25 NM

   f. Initial solo cross-country flight

   g. Solo cross-country flight

    h. Repeated solo cross-country flights not more than 50 NM from point of departure

    i. Solo flight in Class B airspace

    j. Solo flight to, from, or at an airport located in Class B airspace

2. What are the required student pilot certificate endorsements and logbook entries required of you as CFI, before allowing a student pilot to conduct their first solo flight? (AC 61-65)

Prior to solo flight a student pilot is required to receive the following:

    a. An endorsement in their logbook for satisfactory completion of pre-solo aeronautical exam.

    b. An endorsement in their logbook for receiving required pre-solo training in make and model of aircraft.

    c. An endorsement on their student pilot certificate for the make and model of aircraft flown.

3. What are the required student pilot certificate endorsements and logbook entries required of you as CFI, before allowing a student pilot to conduct the initial solo cross-country flight? (AC 61-65)

Prior to the initial solo cross-country flight the student is required to receive:

    a. An endorsement in their logbook from the instructor who has given the required solo cross-country training for the make and model of aircraft to be flown.

    b. An endorsement on their student pilot certificate for the specific category to be flown.

    c. An endorsement in their logbook for the specific solo cross-country flight to be flown.

Note: The CFI should also determine the student's solo flight endorsement is current for the make and model to be flown.

4. Your student is making the final preparations for a solo cross-country flight you have instructed them to make. You are not available to review the planning and preparation as required. Can another CFI conduct the review and provide the necessary endorsement? (AC 61-65)

Yes; before each cross-country flight, the student's logbook must be endorsed by an instructor, but this instructor need not be the instructor who provided the initial cross-country instruction to the student.

5. Give examples of additional endorsements you may give as a CFI. (AC 61-65)

Within the limitations of your flight instructor certificate and ratings, and your pilot certificate and ratings, you may give training and endorsements for:

a. Completion of a flight review
b. Completion of a phase of an FAA-sponsored pilot proficiency award program
c. Completion of an instrument proficiency check
d. To act as PIC in a complex airplane
e. To act as PIC in a high-performance airplane
f. To act as PIC in a pressurized aircraft capable of high altitude operations
g. To act as PIC in a tailwheel airplane
h. To act as PIC of an aircraft in solo operation when the pilot does not hold an appropriate category and class rating
i. Retesting after failure of a knowledge or practical test

6. Must the student pilot's certificate be endorsed every 90 days or is this endorsement required only in the student pilot's logbook? (AC 61-65)

The 90-day solo endorsement that is entered in the student's logbook is required every 90 days for continuing solo privileges. The endorsement on the student pilot certificate is a one-time endorsement.

Additional Study Questions:

1. Your student recently soloed in a C-152. The student has expressed interest in transitioning to a C-172 for the remainder of the flight training and the checkride. What actions must you take to allow this student to solo a C-172? (14 CFR 61.87)

2. If a student pilot has their student pilot certificate endorsed to fly more than one model of aircraft in solo flight, is a logbook endorsement required for each one of these models? (14 CFR 61.87) Yes

3. Another CFI has requested you review the preflight planning and preparation of their student for a solo cross-country flight. While conducting the review you notice that the 90-day solo endorsement has expired. Are you allowed to provide another 90-day solo endorsement or must you first fly with that student? (14 CFR 61.93) No!

4. Can another flight instructor endorse a student pilot's certificate for cross-country flight without first having given that student cross-country flight instruction? (14 CFR 61.93) No

5. What distance from the point of departure is a student pilot allowed to fly, if his/her logbook has not been endorsed for cross-country flights? (14 CFR 61.93) 25 nm

6. As a CFI-ASEL with no tailwheel experience, are you allowed to give a tailwheel endorsement?

Takeoffs and Climbs

4

A. Normal Takeoff and Climb

1. Describe a normal takeoff and climb.

A normal takeoff and climb procedure is one in which the airplane
is headed directly into the wind or the wind is very light, and the
takeoff surface is firm with no obstructions along the takeoff path,
and is of sufficient length to permit the airplane to gradually accelerate to normal climbing speed.

2. Discuss the importance of a thorough knowledge of normal takeoff and climb principles and procedures.

A thorough knowledge of takeoff and climb principles, both in
theory and practice, will prove to be of great value throughout the
pilot's career. It often may prevent an attempt to takeoff under
critical conditions that would require performance beyond the
capability of the airplane or skill of the pilot. The takeoff itself,
though relatively simple, often presents the most hazards of any
part of a flight.

3. What are the steps involved in performing a normal takeoff and climb procedure?

Although the takeoff and climb process is one continuous maneuver, it can be divided into three main steps in order to explain the
process:

a. The takeoff roll:

   i. Align the airplane with the runway centerline;

   ii. Apply throttle smoothly and continuously to maximum
allowable power;

   iii. Maintain directional control with rudder; slight rudder pressure will be required to compensate for torque;

   iv. Glance at the engine instruments for any sign of malfunction.

b. The liftoff:

   i. As soon as all flight controls become effective during the
takeoff roll, back pressure should be applied gradually to lift
the nose wheel off of the runway (rotation);

Continued

    ii. Adjust and maintain liftoff attitude for V_X or V_Y;

    iii. Keep the wings level and establish an initial heading.

c. The initial climb after becoming airborne:

    i. Establish pitch attitude for V_X or V_Y as necessary;

    ii. Retrim aircraft for appropriate speed;

    iii. Maintain takeoff power until 500 AGL above surrounding terrain;

    iv. Adjust heading to maintain track of extended runway centerline.

4. What are the standards expected of a student for a normal takeoff and climb? (FAA-S-8081-12A)

The student:

a. Shows knowledge of normal and crosswind takeoff and climb.

b. Positions flight controls for existing conditions.

c. Clears the area, taxies into the takeoff position, and aligns the airplane on the runway center.

d. Advances throttle to takeoff power.

e. Rotates at recommended airspeed and accelerates to V_Y, ± 5 knots*.

f. Retracts landing gear after a positive rate-of-climb is established.

g. Maintains takeoff power to a safe maneuvering altitude, then sets climb power.

h. Maintains directional control and proper wind-drift correction throughout takeoff and climb.

i. Uses noise abatement procedures, as required.

j. Completes appropriate checklist.

  * *Private standard: Establish pitch attitude for V_Y, maintain V_Y +10/-5 knots during climb.*

5. What are some common student errors when performing normal takeoffs and climbs? (FAA-S-8081-6AS)

a. Improper initial positioning of flight controls and wing flaps. No aileron deflection for crosswind; flaps not set as recommended.

b. Improper power application—not applying full power or applying power too quickly or too slowly.

c. Inappropriate removal of hand from throttle. Hand should always remain on throttle during maneuvers such as takeoff, landing, slow flight, etc.

d. Poor directional control.

   i. Not correcting for torque effect.

   ii. Overcorrecting or undercorrecting with rudder.

e. Improper use of ailerons.

   i. In a crosswind, as speed increases aileron deflection should be reduced.

   ii. Allowing the upwind wing to rise causing airplane to skip sideways.

f. Improper pitch attitude during liftoff.

   i. Forcing the airplane off the runway; too much pitch attitude too soon.

   ii. Airplane lifts off at too slow an airspeed causing it to stall back to the runway.

   iii. Allowing the takeoff roll to continue causing the airplane to remain on the runway too long.

g. Failure to establish and maintain proper climb configuration and airspeed.

   i. Not establishing the pitch attitude for best rate-of-climb.

   ii. Not retracting flaps/gear as appropriate.

h. Drift during climb.

   i. Allowing the airplane to drift away from the runway extended centerline.

   ii. Not clearing area directly in front of aircraft during climb.

B. Crosswind Takeoff and Climb

1. Describe a crosswind takeoff and climb.

A crosswind takeoff is a takeoff performed when the wind direction is from other than directly in front of the airplane.

2. Discuss the importance of knowing crosswind takeoff and climb principles and techniques.

While it is usually preferable to take off directly into the wind whenever possible or practical, there will be many instances when circumstances or wisdom will indicate otherwise. Consequently, the pilot must be familiar with the principles and techniques involved in crosswind takeoffs as well as those for normal takeoffs.

3. What are the principles and procedures involved in a crosswind takeoff and climb?

a. Crosswind takeoff roll:

   i. Taxi into takeoff position and check strength and direction of wind.

   ii. Start takeoff roll with full aileron control into the wind.

   iii. As the forward speed of the airplane increases and the crosswind becomes more and more of a relative headwind, the mechanical holding of full aileron into the wind should be reduced.

b. Crosswind liftoff:

   i. If a significant crosswind exists, the main wheels should be held on the ground slightly longer than in a normal takeoff so that a smooth but very definite liftoff can be made.

   ii. As both main wheels leave the runway, adequate drift correction must be maintained by the pilot or the airplane will slowly be carried sideways with the wind.

c. Initial crosswind climb:

   i. If proper correction is being applied, as soon as the airplane becomes airborne it will be slipping into the wind sufficiently to counteract the drifting effect of the wind.

ii. This slipping should be continued until the airplane has climbed well above the ground. At that time the airplane should be headed toward the wind to establish just enough "crab" to counteract the wind and then the wings rolled level.

iii. The climb while in this crab should be continued so as to follow a ground track aligned with the runway direction.

4. What are the standards expected of a student for a crosswind takeoff and climb? (FAA-S-8081-12A)

The student:

a. Shows knowledge of normal and crosswind takeoff and climb.

b. Positions flight controls for existing conditions.

c. Clears the area, taxies into the takeoff position, and aligns the airplane on the runway center.

d. Advances throttle to takeoff power.

e. Rotates at recommended airspeed and accelerates to V_Y, ±5 knots*.

f. Retracts landing gear after a positive rate-of-climb is established.

g. Maintains takeoff power to a safe maneuvering altitude, then sets climb power.

h. Maintains directional control and proper wind-drift correction throughout takeoff and climb.

i. Uses noise abatement procedures, as required.

j. Completes appropriate checklist.

* *Private standard: Establish pitch attitude for V_Y, maintain V_Y +10/-5 knots during climb.*

5. What are some common student errors in performing crosswind takeoffs and climbs? (FAA-S-8081-6AS)

a. Improper initial positioning of flight controls and wing flaps.

    i. If a crosswind exists, full aileron into wind should be applied initially.

    ii. Flaps should be set as recommended by manufacturer.

b. Improper power application.

    i. Not applying full power.

    ii. Applying power too quickly or too slowly.

c. Inappropriate removal of hand from throttle. The hand should always remain on throttle during maneuvers such as takeoff, landing, slow flight, etc.

d. Poor directional control.

    i. Not correcting for torque effect.

    ii. Overcorrecting or undercorrecting with rudder.

e. Improper use of ailerons.

    i. As speed increases aileron deflection should be reduced.

    ii. Allowing the upwind wing to rise causing the airplane to skip sideways.

f. Improper pitch attitude during liftoff.

    i. Forcing the airplane off the runway; too much pitch attitude too soon.

    ii. Airplane lifts off at too slow an airspeed causing it to stall back to the runway.

    iii. Allowing takeoff roll to continue; airplane remains on the runway to long.

g. Failure to establish and maintain proper climb configuration and airspeed.

    i. Not establishing pitch attitude for best rate-of-climb.

    ii. Not retracting flaps/gear as appropriate.

h. Drift during climb.

    i. Allowing airplane to drift away from the runway extended centerline.

    ii. Not clearing area directly in front of aircraft during climb.

C. Short-Field Takeoff and Climb

1. What is the purpose of short-field takeoff and climb procedures?

As the name implies, short-field takeoff procedures are utilized when an airplane must be operated out of an area with either a short runway and/or the available takeoff area is restricted by obstructions.

2. Why is it important that a pilot be familiar with these procedures?

A takeoff and climb from a field where the takeoff area is short or the available takeoff area is restricted by obstructions, requires that the pilot operate the airplane at the limit of its takeoff performance capabilities. To depart such an area safely, the pilot must exercise positive and precise control of the airplane attitude and airspeed so that takeoff and climb performance results in the shortest ground roll and the steepest angle of climb. In order to accomplish a maximum performance takeoff safely, the pilot must be well indoctrinated in the use and effectiveness of best-angle-of-climb speed and best-rate-of-climb speed for the specific make and model of airplane flown.

3. Describe the short-field takeoff and climb procedures.

a. Set flaps as recommended by manufacturer.

b. Taxi onto runway using all available runway length.

c. Momentarily apply brakes while applying maximum allowable power.

d. Check all engine instruments in the green and release brakes.

e. Adjust airplane pitch attitude/angle of attack for minimum drag and maximum acceleration.

f. Accelerate to recommended liftoff airspeed.

g. On liftoff, adjust pitch attitude for V_X (best angle) until obstacles cleared or if no obstacles an altitude at least 50 feet AGL is obtained.

h. Retract flaps and gear (if retractable) when well clear of obstacles and best rate-of-climb has been established.

4. What are the standards expected of a student for short-field takeoff and climb? (FAA-S-8081-12A)

The student:

a. Shows knowledge of short-field takeoff and climb.

b. Positions flight controls and flaps for existing conditions.

c. Clears the area, taxies into position for maximum utilization of available takeoff area.

d. Advances throttle smoothly to takeoff power while holding brakes (or as specified).

e. Rotates at recommended airspeed.

f. Climbs at manufacturer's recommended configuration and airspeed, or in their absence at V_X, +5/-0 knots until obstacle is cleared, or until the airplane is at least 50 feet (20 meters) above the surface.*

g. After clearing obstacle, accelerates to and maintains V_Y, ±5 knots.

h. Retracts landing gear and flaps after a positive rate-of-climb is established (or as specified).

i. Maintains takeoff power to a safe maneuvering altitude, then sets climb power.

j. Maintains directional control and proper wind-drift correction throughout takeoff and climb.

k. Completes appropriate checklists.

* *Private standard: Establish pitch attitude for recommended airspeed or V_X, maintain at airspeed, +10/-5 knots, during climb.*

5. Describe some common errors for students performing short-field takeoff and climb. (FAA-S-8081-6AS)

a. Failure to position the airplane for maximum utilization of available takeoff area.

b. Improper initial positioning of flight controls and wing flaps.

   i. No aileron deflection for crosswind.

   ii. Flaps not set as recommended.

c. Improper power application.

     i. Not applying full power.

     ii. Applying power too quickly or too slowly.

 d. Inappropriate removal of hand from throttle. Hand should always remain on throttle during maneuvers such as takeoff, landing, slow flight, etc.

 e. Poor directional control.

     i. Not correcting for torque effect.

     ii. Overcorrecting or undercorrecting with rudder.

 f. Improper use of brakes.

     i. Failure to hold brakes until full power is developed and engine instruments are checked.

     ii. Failure to remove feet from brakes during takeoff roll.

 g. Improper pitch attitude during liftoff.

     i. Too much pitch attitude too soon.

     ii. Forcing the airplane off the runway; airplane lifts off at too slow an airspeed causing it to stall back to the runway.

     iii. Allowing takeoff roll to continue; airplane remains on the runway too long.

 h. Failure to establish and maintain proper climb configuration and airspeed.

     i. Retracting flaps/landing gear before clear of obstacle.

     ii. Not maintaining best angle of climb prior to flaps/gear retraction.

 i. Drift during climb.

     i. Allowing airplane to drift away from runway extended centerline.

     ii. Not clearing area directly in front of aircraft during climb.

D. Soft-Field Takeoff and Climb

1. What is the purpose of soft-field takeoff and climb procedures?

Soft-field takeoff and climb procedures are utilized when operating an airplane off of an unimproved surface such as grass, soft sand, mud, snow, or rough terrain, etc.

2. Why is it important that a pilot be familiar with these procedures?

Soft-field takeoffs and climbs require the use of operational techniques for getting the airplane airborne as quickly as possible. Soft surfaces such as tall grass, soft sand, mud, snow, etc. usually retard the airplane's acceleration during the takeoff roll so much that adequate takeoff speed might not be attained if normal techniques were employed. Soft-field takeoff and climb procedures are also useful when operating an airplane off of a rough field where it is advisable to get the airplane off the ground as soon as possible to avoid damaging the landing gear.

3. Describe the soft-field takeoff and climb procedures.

a. Wing flaps should be lowered prior to starting the takeoff roll (if recommended by the manufacturer).

b. Taxi the airplane at as fast a speed as possible, consistent with safety and surface conditions. Avoid making sharp turns, using brakes, and any other action which might bog the airplane down.

c. The airplane should be kept in continuous motion with sufficient power while lining up for the takeoff roll.

d. As the airplane is aligned, apply power smoothly to maximum allowable power.

e. As the airplane accelerates, enough elevator back pressure should be applied to reduce the weight supported by the nose wheel.

f. Maintain a nose-high attitude throughout the takeoff run sufficient to relieve the main gear of progressively more and more weight. This will minimize drag caused by surface irregularities or adhesion.

g. As the airplane becomes airborne lower pitch attitude slightly to gain additional airspeed while in ground effect.

h. Accelerate to V_X with obstacle or V_Y without obstacle before leaving ground effect.

i. Continue climb at V_X or V_Y as appropriate.

j. Retract the wing flaps and/or landing gear when clear of obstacles.

4. What are the standards expected of a student for soft-field takeoff and climb? (FAA-S-8081-12A)

The student:

a. Shows knowledge of soft-field takeoff and climb.

b. Positions flight controls and flaps for existing conditions to maximize lift as quickly as possible.

c. Clears the area, taxies onto the takeoff surface at a speed consistent with safety and aligns the airplane without stopping while advancing the throttle smoothly to takeoff power.

d. Establishes and maintains a pitch attitude that will transfer the weight of the airplane from the wheels to the wings.

e. Remains in ground effect after takeoff while accelerating to V_X or V_Y, as required.

f. Maintains V_Y, ±5 knots*.

g. Retracts landing gear and flaps after a positive rate-of-climb is established (or as specified).

h. Maintains takeoff power to a safe maneuvering altitude, then sets climb power.

i. Maintains directional control and proper wind-drift correction throughout takeoff and climb.

j. Completes appropriate checklists.

* *Private standard: Establish pitch attitude for V_Y, maintain V_Y, +10/-5 knots, during climb.*

5. What are some common errors for students performing soft-field takeoff and climb? (FAA-S-8081-6AS)

a. Improper initial positioning of flight controls and wing flaps.

    i. No aileron deflection for crosswind.

    ii. Flaps not set as recommended.

b. Hazards of allowing the airplane to stop on the takeoff surface prior to initiating takeoff—aircraft may not have enough power to begin takeoff roll again.

Continued

   c. Improper power application.

      i. Not applying full power.

      ii. Applying power too quickly or too slowly.

   d. Inappropriate removal of hand from throttle. Hand should always remain on throttle during maneuvers such as takeoff, landing, slow flight, etc.

   e. Poor directional control.

      i. Not correcting for torque effect.

      ii. Overcorrecting or undercorrecting with rudder.

   f. Improper use of brakes. Use of brakes should not be required in soft field operations.

   g. Improper pitch attitude during liftoff.

      i. Not applying full up elevator during the initial takeoff roll delaying liftoff.

      ii. Not reducing full up elevator during the takeoff roll delaying liftoff.

      iii. Allowing airplane to liftoff and then stall back onto runway.

   h. Hazards of settling back to takeoff surface after becoming airborne. Excessive loads may be inflicted on the landing gear with possible loss of control of aircraft.

   i. Failure to establish and maintain proper climb configuration and airspeed.

      i. Not remaining in ground effect while accelerating to V_X or V_Y as appropriate.

      ii. Not maintaining V_X until clear of obstacle.

      iii. Retracting flaps before clear of obstacle.

   j. Drift during climb.

      i. Allowing airplane to drift away from runway extended centerline.

      ii. Not clearing areas directly in front of aircraft during climb.

Fundamentals
of Flight

5

A. Straight-and-Level Flight

1. What is straight-and-level flight?

Straight-and-level flight is just what the name implies—flight in which a constant heading and altitude are maintained. It is accomplished by making immediate corrections for deviations in direction and altitude from unintentional slight turns, descents, and climbs.

2. What is the objective in learning straight-and-level flight maneuvers?

The objective in all the basic maneuvers is to learn the proper use of the controls for maneuvering the airplane, to attain the proper attitude in relation to the horizon by use of inside and outside references (the integrated flight instructor method), and to emphasize the importance of dividing attention and constantly checking all reference points.

3. How is straight-and-level flight achieved?

The pitch attitude for level flight (constant altitude) is usually obtained by selecting some portion of the airplane's nose as a reference point, and then keeping that point in a fixed position relative to the horizon. That position should be cross-checked occasionally against the altimeter and attitude indicator to determine whether or not the pitch attitude is correct.

To achieve straight flight (laterally level flight) the pilot should select two or more outside visual reference points directly ahead of the airplane (such as fields, towns, lakes), to form points along an imaginary line and keep the airplane's nose headed along that line. While using these references, an occasional check of the heading indicator and attitude indicator should be made to determine that the airplane is actually maintaining flight in a constant direction.

4. What are some common student errors in the performance of straight-and-level flight? (FAA-S-8081-6AS)

a. Failure to cross-check and correctly interpret outside and instrument references. Fixating on instruments inside instead of utilizing a combination of inside and outside references.

b. Application of control movements rather than pressures— use of jerky control movements instead of smooth control pressures.

c. Uncoordinated use of flight controls—not applying right rudder to compensate for torque effect in straight-and-level flight.

d. Faulty trim technique:

   i. Failure to trim the aircraft.
   ii. Excessive use of trim.
   iii. Using trim as a primary flight control.
   iv. Trimming the aircraft before establishing pitch attitude and power setting.

B. Level Turns

1. Describe a level turn.

A turn is a basic flight maneuver used to change or return to a desired heading. It involves close coordination of all three flight controls—aileron, rudder, and elevator.

2. Discuss the importance of the level turn, as a basic maneuver.

Since turns are a part of almost every other flight maneuver, it is important that the pilot thoroughly understand the factors involved and learn to perform them well.

3. What are the steps involved in performing a level turn?

a. Roll into the banked attitude by coordinated use of aileron and rudder in the direction of turn.

b. When the desired angle of bank is obtained, neutralize the ailerons and rudder to maintain bank.

c. Back pressure must be applied in the turn to compensate for the loss of vertical lift and to maintain altitude.

d. Roll out of the turn by applying coordinated aileron and rudder pressure in the opposite direction of the turn until level attitude is reached. As the angle of bank is decreased the elevator should be released smoothly as necessary to maintain altitude.

4. What are some common student errors in performing level turns? (FAA-S-8081-6AS)

a. Failure to cross-check and correctly interpret outside and instrument references. Fixating on instruments inside instead of utilizing a combination of inside and outside references.

b. Application of control movements rather than pressures— use of jerky control movements instead of smooth control pressures.

c. Uncoordinated use of flight controls.

   i. Initially too much bank or rudder when establishing the turn.

   ii. Slipping or skidding during the turn.

d. Faulty altitude and bank control.

   i. Excessive or insufficient back pressure resulting in a gain or loss of altitude.

   ii. Bank angle varies due to lack of division of attention between inside and outside references.

C. Straight Climbs and Climbing Turns

1. What is a straight climb? A climbing turn?

A straight climb is one in which the airplane gains altitude while traveling straight ahead. It is a basic flight maneuver in which an increase in both the pitch attitude and power result in a gain in altitude. Climbing turns are those in which the airplane gains altitude while turning.

2. What is the objective in learning the straight climb/ climbing turn maneuver?

The objective of this maneuver, as with the other basic flight maneuvers, is to learn the proper use of the controls for maneuvering the airplane, to attain the proper attitude in relation to the horizon by use of inside and outside references, and to emphasize the importance of dividing attention and constantly checking all references points.

3. Describe how to achieve the straight climb/climbing turn maneuver.

a. Establish a climb by applying back pressure on the elevator to increase pitch attitude. Simultaneously establish the desired bank angle if performing a climbing turn.

b. Apply full power and establish a pitch attitude for the climbing airspeed (V_Y).

c. Cross-check the airspeed indicator with the position of the airplane's nose in relation to the horizon as well as the attitude indicator.

d. Trim the aircraft for this attitude/airspeed.

e. Utilize right rudder to correct for torque effect.

f. Maintain a constant heading by cross-checking visual references as well as instrument references. Maintain wings level while cross-checking heading indicator, attitude indicator, turn coordinator. If performing a climbing turn, maintain the desired bank angle by cross-checking visual references as well as instrument references.

4. What are some common student errors in performing straight climbs/climbing turns? (FAA-S-8081-6AS)

a. Failure to cross-check and correctly interpret outside and instrument references—fixating on instruments inside instead of utilizing a combination of inside and outside references.

b. Application of control movements rather than pressures— use of jerky control movements instead of smooth control pressures.

   c. Uncoordinated use of flight controls.

      i. Not compensating for torque effect in climb.

      ii. Slipping or skidding during the turn.

   d. Faulty trim technique.

      i. Failure to trim the aircraft.

      ii. Excessive use of trim.

      iii. Using trim as a primary flight control.

      iv. Trimming the aircraft before establishing pitch attitude and power setting.

D. Straight Descents and Descending Turns

1. Describe the straight descent and descending turn maneuvers.

A descent or a glide is a basic maneuver in which the airplane is losing altitude in a controlled descent with little or no engine power; forward motion is maintained by gravity pulling the airplane along an inclined path, and the descent rate is controlled by the pilot balancing the forces of gravity and lift. Descending turns are those in which the airplane loses altitude while turning.

2. Explain the objective in learning straight descents/descending turns, and why these maneuvers are especially important.

Although power-off descents (glides) and descending turns are directly related to the practice of power-off accuracy landings, they have a specific operational purpose in normal landing procedures, and forced landings after engine failure. Therefore, it is necessary that they be performed more subconsciously than other maneuvers because, most of the time during their execution, the pilot will be giving full attention to details other than the mechanics of performing the maneuver. Since power-off descents and descending turns are performed relatively close to the ground, accuracy of their execution and the formation of proper technique and habits are of special importance.

3. How is a straight descent or descending turn maneuver performed?

a. Apply carburetor heat and reduce power to idle.

b. Maintain a level pitch attitude to reduce airspeed to the recommended glide speed.

c. Establish the desired bank angle if performing a descending turn.

d. Allow the pitch attitude to decrease as necessary to maintain best glide speed.

e. When airspeed is stabilized, the aircraft should be retrimmed.

4. What are some common errors in performing straight descents/descending turns? (FAA-S-8081-6AS)

a. Failure to cross-check and correctly interpret outside and instrument references—fixating on instruments inside instead of utilizing a combination of inside and outside references.

b. Application of control movements rather than pressures—use of jerky control movements instead of smooth control pressures.

c. Uncoordinated use of flight controls.

   i. Initially too much bank or rudder when establishing the turn.
   ii. Slipping or skidding during the turn.

d. Faulty trim technique.

   i. Failure to trim the aircraft.
   ii. Excessive use of trim.
   iii. Using trim as a primary flight control.
   iv. Trimming aircraft before establishing pitch attitude and power setting.

e. Failure to clear engine and use carburetor heat, as appropriate.

E. Practical Test Standards for the Basic Maneuvers

1. What are the practical test standards for teaching basic flight maneuvers? (FAA-S-8081-6AS)

The flight instructor exhibits instructional knowledge of the elements of straight-and-level flight, level turns, straight climbs/climbing turns, and straight descents and descending turns by describing:

a. The effect and use of flight controls.

b. The integrated flight instruction method.

c. Outside and instrument references used for pitch, bank, and power control; the cross-check and interpretation of those references; and the control technique used.

d. Trim technique.

e. Methods of overcoming tenseness and overcontrolling.

2. What recommended standards should a flight instructor use to evaluate a student's performance of the basic flight maneuvers?

The student:

a. Maintains desired heading plus or minus 10°.
b. Maintains desired altitude plus or minus 100 feet.
c. Maintains desired airspeed plus or minus 10 knots.
d. Maintains desired bank angle.

Note: The four basic tasks in this chapter, when conducted visually and not by reference to instruments, are not specified in the Private or Commercial practical test standards. The above standards are recommended for evaluation purposes only.

Stalls, Spins, and Maneuvers During Slow Flight

6

A. Power-On Stalls

1. What is a power-on stall?

Power-on stalls are, as the name implies, stalls in which full power is being developed as the aircraft stalls. They are intended to simulate the characteristics of an airplane that has stalled in a takeoff and departure configuration.

2. Explain the objective in learning the power-on stall.

The objectives in performing stalls are to familiarize the pilot with the conditions that produce stalls, to assist in recognizing a takeoff and departure stall, and to develop the habit of taking prompt preventive or corrective action.

3. What are the steps involved in performing a power-on stall?

a. Perform clearing turns.

b. Establish the heading and altitude (recovery by 1,500 feet AGL).

c. Establish the takeoff or departure configuration.

d. Slow the airplane to normal liftoff airspeed.

e. Apply takeoff power for a takeoff stall, or the recommended climb power for a departure stall.

f. Establish a climb attitude.

g. After the climb attitude is established, the nose should be brought smoothly upward to an attitude obviously impossible for the airplane to maintain and held at that attitude until a full stall occurs.

h. Recovery should be accomplished by immediately reducing the pitch attitude/angle of attack, applying maximum power (not necessary in a takeoff stall) and maintaining directional control through coordinated use of controls.

i. Control any yawing tendency with rudder.

Continued

j. Utilize ailerons to level wings as soon as possible.

k. As airspeed approaches V_X establish climb attitude to maintain V_X and to establish a positive rate climb.

l. Return to cruise flight.

4. What are the standards expected of a student for performing a power-on stall? (FAA-S-8081-12A)

The student:

a. Shows knowledge of aerodynamic factors associated with power-on stalls and how this relates to actual takeoff and departure situations.

b. Selects an entry altitude that allows the task to be completed no lower than 1,500 feet (460 meters) AGL or the manufacturer's recommended altitude, whichever is higher.

c. Establishes takeoff configuration and slows airplane to normal lift-off speed.

d. Sets power to manufacturer's recommended power-on stall power setting while establishing climb attitude (in the absence of a manufacturer—recommended power setting, use no less than approximately 55-60 percent of full power as a guideline).

e. Maintains the specified heading ±10°, in straight flight; maintains a specified angle of bank, not to exceed a 20° angle of bank, ±10°, in turning flight.

f. Recognizes and announces onset of stall by identifying the first aerodynamic buffeting or decay of control effectiveness.

g. Recovers promptly as stall occurs, by simultaneously decreasing pitch attitude, increasing power and leveling the wings, with minimum loss of altitude.

h. Retracts flaps (if applicable) and landing gear after a positive rate of climb is established.

i. Returns to the altitude, heading, and airspeed specified by the examiner.

Note: In some high performance airplanes, the power setting may have to be reduced below the practical test standards guideline power setting to prevent excessively high pitch attitudes (greater than 30° nose up).

5. What are some common student errors in a performance of a power-on stall? (FAA-S-8081-6AS)

a. Failure to establish the specified landing gear and flap configuration prior to entry.

b. Improper pitch, heading, and bank control during straight ahead stalls.

   i. Not reducing power initially to slow the airplane to a typical takeoff and departure airspeed.

   ii. Increasing the pitch attitude too much, too quickly, resulting in an excessively steep nose-up attitude and an unrealistic situation.

c. Improper pitch and bank control during turning stalls.

   i. Not reducing power initially to slow the airplane to a typical takeoff and departure airspeed.

   ii. Increasing the pitch attitude too much, too quickly, resulting in an excessively steep nose-up attitude and an unrealistic situation.

   iii. Not maintaining the specified bank angle.

d. Rough or uncoordinated control technique.

   i. Not utilizing rudder to assist in maintaining initial directional control.

   ii. All aileron and no rudder will only aggravate the situation, especially before the wings have had time to regain sufficient airflow.

e. Failure to recognize the first indications of a stall—not recognizing initial buffeting and lack of control effectiveness.

f. Failure to achieve a stall. Not increasing pitch attitude high enough to induce a stall or initiating recovery before stall occurs.

g. Improper torque correction—not correcting for torque effect with right rudder.

h. Poor stall recognition and delayed recovery—not reducing back pressure after stall has occurred.

i. Excessive altitude loss or excessive airspeed during recovery—pitch attitude is reduced to an excessive nose down attitude, or is maintained in a nose down attitude longer than necessary.

j. Secondary stall during recovery—student hastens recovery by increasing pitch attitude too quickly.

B. Power-Off Stalls

1. What is a power-off stall?

A power-off stall is a maneuver designed to simulate an accidental stall occurring during a segment of a landing approach. Airplanes equipped with flaps and/or retractable gear will be in the landing configuration. Power-off stalls are practiced from straight-ahead flight as well as from moderately banked turns to simulate an accidental stall during a turn from base leg to final.

2. Discuss the advantages of learning the power-off stall.

Power-off stalls are practiced to show what could happen if the controls are improperly used during a turn from the base leg to the final approach. The power-off straight ahead stall simulates the attitude and flight characteristics of a particular airplane during the final approach and landing. During the practice of intentional stalls, the real objective is not to learn how to stall the airplane but to learn how to recognize an incipient stall and take prompt corrective action. Practice of stalls is important because it simulates stall conditions that could occur during normal flight maneuvers.

3. How is a power-off stall performed?

a. Perform clearing turns.

b. Establish heading and altitude (recovery by 1,500 feet AGL).

c. Extend landing gear (if applicable).

d. Carburetor heat ON.

e. Reduce power and maintain back pressure to slow aircraft to flap operating speed.

f. Extend approach flaps.

g. Reduce power to idle and establish approach airspeed.

h. Bring the nose smoothly upward until the full stall occurs.

i. Immediately reduce angle of attack to regain flying speed.

j. Simultaneously apply full power (carburetor heat off).

k. Retract flaps incrementally.

l. Control any yawing tendency with rudder.

m. Utilize ailerons to level wings as soon as possible.

n. As airspeed approaches V_X establish climb attitude to maintain V_X and to establish a positive rate climb.

o. Return to cruise flight.

4. What are the standards expected of a student for performing a power-off stall? (FAA-S-8081-12A)

The student:

a. Shows knowledge of aerodynamic factors associated with power-off stalls and how this relates to actual approach and landing situations.

b. Selects an entry altitude that allows the task to be completed no lower than 1,500 feet (460 meters) AGL or the manufacturer's recommended altitude, whichever is higher.

c. Establishes stabilized descent, in approach or landing configuration, as specified by examiner.

d. Transitions smoothly from approach or landing attitude to pitch attitude that will induce a stall.

e. Maintains specified heading ±10° in straight flight; maintains a specified angle of bank not to exceed 30°, +0/-10°, in turning flight, while inducing a stall.

f. Recognizes and announces onset of the stall by identifying the first aerodynamic buffeting or decay of control effectiveness.

g. Recovers promptly as stall occurs by simultaneously decreasing pitch attitude, increasing power and leveling the wings, with minimum loss of altitude.

h. Retracts flaps to recommended setting; retracts landing gear after a positive rate-of-climb is established.

i. Accelerates to V_X or V_Y speed before final flap retraction (or as recommended).

j. Returns to altitude, heading, and airspeed specified by examiner.

5. List some common student errors in performing the power-off stall. (FAA-S-8081-6AS)

a. Failure to establish the specified landing gear and flap configuration prior to entry.

b. Improper pitch and bank control during straight ahead and turning stalls.

   i. Not reducing power initially to slow the airplane to a typical takeoff and departure airspeed.

   ii. Increasing the pitch attitude too much, too quickly, resulting in an excessively steep nose-up attitude and an unrealistic situation.

   iii. Not maintaining the specified bank angle.

c. Rough or uncoordinated control technique.

   i. Not utilizing rudder to assist in maintaining initial directional control.

   ii. All aileron and no rudder will only aggravate the situation, especially before the wings have had time to regain sufficient airflow.

d. Failure to recognize the first indications of a stall—not recognizing the initial buffeting and lack of control effectiveness.

e. Failure to achieve a stall. Not increasing the pitch attitude high enough to induce a stall or initiating recovery before the stall occurs.

f. Improper torque correction. Not correcting for torque effect with right rudder.

g. Poor stall recognition and delayed recovery. Not reducing back pressure after the stall has occurred.

h. Excessive altitude loss or excessive airspeed during recovery. Pitch attitude is reduced to an excessive nose down attitude, or is maintained in a nose down attitude longer than necessary.

i. Secondary stall during recovery. Student hastens recovery by increasing pitch attitude too quickly.

C. Crossed-Control Stalls

1. What happens in a crossed-control stall?

A crossed-control stall occurs when the pilot allows the aircraft to be flown in uncoordinated flight with the flight controls crossed—that is, aileron pressure applied in one direction and rudder pressure in the opposite direction. If excessive back pressure is applied, a crossed-control stall may result.

2. Explain why the flight instructor should demonstrate the crossed-control stall to the student.

The objective of this demonstration maneuver is to show the effect of improper control technique and to emphasize the importance of using coordinated control pressures whenever making turns. This type of stall is most likely to occur during a poorly planned and executed base to final approach turn and often is the result of overshooting the centerline of the runway during that turn.

3. How is the crossed-control stall demonstrated?

a. Perform clearing turns.

b. Establish heading and altitude (recovery by 1,500 feet AGL).

c. Perform a checklist (GUMPS).

d. Reduce power.

e. Maintain altitude until airspeed approaches normal glide speed.

f. Retrim aircraft.

g. Roll into a medium-banked turn.

h. During turn excessive rudder pressure should be applied in direction of turn but the bank held constant by applying opposite aileron pressure.

i. Increase back elevator pressure to keep nose from lowering.

j. Control pressures should be increased until airplane stalls.

k. When stall occurs, recover by releasing control pressures and increasing power as necessary.

Continued

l. Control any yawing tendency with rudder.

m. Utilize ailerons to level wings as soon as possible.

n. As airspeed approaches V_X establish climb attitude to maintain V_X and to establish a positive rate climb.

o. Return to cruise flight.

4. What are some common errors related to crossed-control stalls? (FAA-S-8081-6AS)

a. Failure to establish selected configuration prior to entry.

b. Failure to establish a crossed-control turn and stall condition that will adequately demonstrate the hazards of a crossed-control stall.

   i. Not reducing power initially to slow the airplane to a typical approach speed.

   ii. Not increasing crossed-control pressures enough to induce a stall.

   iii. Not increasing back elevator pressure enough to induce a stall.

c. Improper or inadequate demonstration of the recognition of and recovery from a crossed-control stall.

d. Failure to present simulated student instruction that adequately emphasizes the hazards of a crossed-control condition in a gliding or reduced airspeed condition. Not explaining the "what, why, and how" of crossed-control stalls adequately.

Practical Test Standards (FAA-S-8081-6AS)

The following is an excerpt from the flight instructor Practical Test Standards set by the FAA for demonstration of crossed-control stalls:

The applicant exhibits instructional knowledge of the elements of crossed-control stalls, with the landing gear extended, by describing the:

a. Aerodynamics of crossed-control stalls.

b. Effects of crossed controls in gliding or reduced airspeed descending turns.

c. Hazards of crossed controls in a base leg to final approach turn.

d. Entry technique and minimum entry altitude.

e. Recognition of crossed-control stalls.

f. Flight situations where unintentional crossed-control stalls may occur.

g. Recovery technique and minimum recovery altitude.

D. Elevator Trim Stalls

1. Describe an elevator trim stall.

An elevator trim stall is a stall that occurs when full power is applied to an airplane configured with excessive nose-up trim. Positive control of the airplane is not maintained resulting in a stall. These type of stalls usually occur during a go-around procedure from a normal landing approach or a simulated forced landing approach, or immediately after takeoff.

2. Why should a flight instructor demonstrate an elevator trim stall?

This maneuver shows what can happen when full power is applied for a go-around and positive control of the airplane is not maintained. The objective of this maneuver is to show the importance of making smooth power applications, overcoming strong trim forces and maintaining positive control of the airplane to hold safe flight attitudes, and using proper and timely trim techniques.

3. How is an elevator trim stall demonstrated?

a. Establish a minimum safe altitude (recovery by 1,500 feet AGL).

b. Perform clearing turns.

c. Slowly retard the throttle and extend landing gear if retractable.

d. Extend one-half to full flaps.

e. Close throttle.

f. Maintain altitude until airspeed approaches normal glide speed.

Continued

g. When normal glide is established, the airplane should be retrimmed just as would be done during a normal landing approach.

h. Advance throttle to maximum power as in a go-around procedure. The combined forces of thrust, torque, and back elevator trim will tend to make the nose rise sharply and turn to the left. To demonstrate what could occur if positive control of the airplane were not maintained, no immediate attempt should be made to correct these forces.

i. When a stall is imminent, forward pressure must be applied to return the airplane to normal climbing attitude.

j. Trim should then be adjusted to relieve the heavy control pressures and the normal go-around and level-off procedures should be completed.

4. What are some common errors associated with elevator trim stalls? (FAA-S-8081-6AS)

a. Failure to establish selected configuration prior to entry.

b. Failure to establish the thrust, torque, and up-elevator trim conditions that will result in a realistic demonstration.

   i. Not establishing a final approach configuration.

   ii. Not applying maximum power, as in a go-around situation.

c. Improper or inadequate demonstration of the recognition of and recovery from an elevator trim stall.

   i. Not allowing the pitch attitude to increase above the normal climbing attitude.

   ii. Reducing power during recovery; not maintaining control of aircraft while retrimming and retracting flaps.

d. Failure to present simulated student instruction that adequately emphasizes the hazards of poor correction for torque and up elevator trim during go-arounds and other maneuvers. Not explaining the "what, why, and how" of elevator trim stalls adequately.

Practical Test Standards (FAA-S-8081-6AS)

The following is an excerpt from the flight instructor Practical Test Standards set by the FAA for demonstration of the elevator trim stall:

The applicant exhibits instructional knowledge of the elements of elevator trim stalls, in selected landing gear and flap configurations, by describing the:

    a. Aerodynamics of elevator trim stalls.

    b. Hazards of inadequate control pressures to compensate for thrust, torque, and up-elevator trim during go-arounds and other related maneuvers.

    c. Entry technique and minimum entry altitude.

    d. Recognition of elevator trim stalls.

    e. Importance of recovering from an elevator trim stall immediately upon recognition.

    f. Flight situations where elevator trim stalls occur.

    g. Recovery technique and minimum recovery altitude.

E. Secondary Stalls

1. What is a secondary stall?

This stall is called a secondary stall since it may occur after a recovery from a preceding primary stall. It is caused by attempting to hasten the completion of a stall recovery before the airplane has regained sufficient flying speed.

2. What should the pilot be aware of regarding secondary stalls?

This stall usually occurs when the pilot becomes too anxious in returning to straight-and-level flight after a stall or spin recovery. Knowledge and proficiency in this maneuver will assist a pilot in avoiding secondary stalls.

3. How should a flight instructor demonstrate a secondary stall?

Secondary stalls can be demonstrated during the recovery phase of any of the basic stalls. The secondary stall can be induced by simply pulling the nose up more rapidly than necessary during the recovery from a full stall.

4. What are some common errors associated with secondary stalls? (FAA-S-8081-6AS)

a. Failure to establish selected configuration prior to entry.

b. Improper or inadequate demonstration of the recognition of and recovery from a secondary stall.

c. Failure to establish a condition that will cause a secondary stall. Not applying sufficient back pressure to induce a secondary stall.

d. Failure to present simulated student instruction that adequately emphasizes the hazards of poor technique in recovering from a primary stall. Not explaining the "what, why, and how" of secondary stalls adequately.

Practical Test Standards (FAA-S-8081-6AS)

The following is an excerpt from the flight instructor Practical Test Standards set by the FAA for secondary stalls:

The applicant exhibits instructional knowledge of the elements of secondary stalls, in selected landing gear and flap configurations, by describing the:

a. Aerodynamics of secondary stalls.

b. Flight situations where secondary stalls may occur.

c. Hazards of secondary stalls during normal stall or spin recovery.

d. Entry technique and minimum entry altitude.

e. Recognition of a secondary stall.

f. Recovery technique and minimum recovery altitude.

F. Spins

1. Describe a spin.

A spin may be described as an aggravated stall that results in what is termed "autorotation" wherein the airplane follows a corkscrew path in a downward direction. Both wings are in a stalled condition but one wing continues to produce some lift resulting in a roll. The airplane is forced downward by gravity, wallowing and yawing in a spiral path.

2. Why is it important for the flight instructor to demonstrate spins and spin recovery?

Fear of and aversion to spins are deeply rooted in the public's mind and many pilots have an unconscious aversion to them. If one learns the cause of a spin and the proper techniques to prevent and/ or recover from the spin, mental anxiety and many causes of unintentional spins may be removed.

3. How should spins and spin recovery be demonstrated?

Entry:

a. Establish the appropriate altitude; recommended minimum altitude for recovery—3,500 feet AGL.

b. Perform clearing turns.

c. Apply carburetor heat and reduce the throttle to idle.

d. Configure the aircraft for a power-off stall (no flaps).

e. As the airplane approaches a stall, smoothly apply full rudder in the direction of the desired spin rotation and continue to apply back elevator to the limit of travel. The ailerons should be neutral.

f. Maintain full rudder deflection and elevator back pressure throughout the spin.

g. Allow the spin to develop (approximately 2 to 3 rotations).

Continued

Recovery:

a. Close the throttle (if not already accomplished).

b. Neutralize the ailerons.

c. Apply full opposite rudder.

d. Briskly move the elevator control forward to approximately the neutral position. (Some aircraft require merely a relaxation of back pressure; others require full forward elevator pressure.)

e. Once the stall is broken the spinning will stop. Neutralize the rudder when the spinning stops.

f. When the rudder is neutralized, gradually apply enough aft elevator pressure to return to level flight.

4. What are some common errors related to spin demonstration? (FAA-S-8081-6AS)

a. Failure to establish proper configuration prior to spin entry. Not establishing configuration recommended by manufacturer for intentional spins.

b. Failure to achieve and maintain a full stall during spin entry. Not maintaining elevator back pressure after entering the spin resulting in a steep spiral.

c. Failure to close throttle when a spin entry is achieved. Not closing throttle resulting in excessive loss of altitude.

d. Failure to recognize the indications of an imminent, unintentional spin. Uncoordinated flight combined with a fully stalled condition and use of incorrect aileron and rudder application for recovery are contributing factors to unintentional spins.

e. Improper use of flight controls during spin entry, rotation or recovery.

    i.   During entry not applying full rudder in direction of spin.

    ii. During rotation not applying full elevator back pressure, full rudder and neutralized ailerons.

    iii. During recovery, not applying full opposite rudder to stop rotation.

    iv. Not applying sufficient forward elevator to break the stalled condition.

    v.   Not utilizing coordinated flight controls during recovery.

f. Disorientation during a spin. Loss of orientation with the outside reference point used to determine number of rotations.

g. Failure to distinguish between a high speed spiral and a spin. Not recognizing a high airspeed (increasing), high rate of descent (increasing), steep spiral condition; the nose of aircraft will not be as low in a steep spiral.

h. Excessive speed or accelerated stall during recovery.

   i. After spin recovery, being too cautious in pulling out of dive resulting in excessive airspeed.

   ii. Applying too much back pressure when recovering, resulting in secondary stall.

i. Failure to recover with minimum loss of altitude.

   i. Not utilizing correct recovery procedures.

   ii. Hesitation in applying necessary control applications.

j. Hazards of attempting to spin an airplane not approved for spins—assuming all airplanes are capable of recovery from intentional spins.

Practical Test Standards (FAA-S-8081-6AS)

The following is an excerpt from the flight instructor Practical Test Standards set by the FAA for spin aerodynamics and demonstration:

Exhibits instructional knowledge of the elements of spins by describing the:

a. Aerodynamics of spins.

b. Airplanes approved for the spin maneuver based on airworthiness category and type certificate.

c. Relationship of various factors such as configuration, weight, center of gravity, and control coordination to spins.

d. Flight situations where unintentional spins may occur.

e. How to recognize and recover from imminent, unintentional spins.

f. Entry technique and minimum entry altitude for intentional spins.

g. Control technique to maintain a stabilized spin.

Continued

h. Orientation during a spin.

i. Recovery technique and minimum recovery altitude for intentional spins.

j. Anxiety factors associated with spin instruction.

G. Maneuvering During Slow Flight

1. What is slow flight?

Slow flight is flight at an airspeed at which any further increase in angle of attack, load factor, or reduction of power will cause an immediate stall. This critical airspeed will depend upon various circumstances, such as gross weight and CG location of the airplane, maneuvering loads imposed by turns and pullups, and the existing density altitude. This maneuver demonstrates the flight characteristics and degree of controllability of an airplane at its minimum flying speed.

2. What is the objective in learning how to operate an airplane at slow flight?

The objective of maneuvering at slow flight is to develop the pilot's sense of feel and ability to use the controls correctly, and to improve proficiency in performing maneuvers in which very low airspeeds are required. The ability to determine the characteristic control responses of an airplane at different airspeeds is of great importance to pilots. They must develop this awareness in order to avoid stalls in airplanes they may fly at the slower airspeeds which are characteristic of takeoffs, climbs, and landing approaches.

3. Describe the procedure for operation at slow flight.

a. Perform clearing turns.

b. Perform a pre-maneuver checklist (GUMPS).

c. Establish a specific heading and altitude (no lower than 1,500 feet AGL).

d. Reduce power from cruise to slow airplane to gear and/or flap operating range.

e. Extend gear (if retractable); extend flaps and adjust pitch attitude to maintain altitude. Retrim aircraft.

f. As the airspeed approaches V_{SO}, utilize power to control altitude and pitch to control airspeed (area of reverse command).

g. Continually cross-check the heading indicator, altimeter, airspeed indicator and vertical airspeed indicator, as well as outside references, to ensure accurate control is maintained.

h. Right rudder should be applied to correct for left turning tendencies.

i. Establish left and right turns (15° bank), climbs and descents while in slow flight.

Recovery:

a. Apply full power.
b. Reduce flaps 10 degrees at a time.
c. Maintain heading and altitude.
d. As airspeed increases, retract gear and any remaining flaps.
e. Retrim aircraft for cruise flight.

4. What are the standards expected of a student for maneuvering during slow flight? (FAA-S-8081-12A)

The student:

a. Shows knowledge of flight characteristics and controllability associated with maneuvering during slow flight.

b. Selects an entry altitude that will allow the task to be completed no lower than 1,500 feet (460 meters) AGL or the manufacturer's recommended altitude, whichever is higher.

c. Stabilizes and maintains airspeed at 1.2 V_{S1}, ±5 knots.

   Private PTS: Stabilizes the airspeed at 1.2 V_{S1}, +10/-5 knots.

d. Establishes straight-and-level flight and level turns, with gear and flaps selected as specified by examiner.

e. Maintains specified altitude, ±50 feet (20 meters).

   Private PTS: Maintains the specified altitude, ±100 feet (30 meters).

Continued

f. Maintains specified heading during straight flight ±10°.

Private PTS: Maintains the specified heading. ±10 degrees, during straight flight.

g. Maintains specified bank angle ±10° during turning flight.

Private PTS: Maintains the specified angle of bank, not to exceed 30 degrees in level flight, +0/-10 degrees; maintains the specified angle of bank, not to exceed 20 degrees in climbing or descending flight, +0/-10 degrees; rolls out on the specified heading, ±10 degrees; and levels off from climbs and descents within ±100 feet (30 meters).

h. Rolls out on specified headings ±10°.

i. Divides attention between airplane control and orientation.

5. What are some common student errors in flight at slow flight that a flight instructor should be aware of? (FAA-S-8081-6AS)

a. Failure to establish the specified configuration.

b. Improper entry technique:

i. Difficulty in transition from cruise flight to slow flight.

ii. Not increasing back pressure as power is reduced.

iii. Increasing back pressure too quickly when power is reduced.

iv. As airspeed slows, failure to apply pitch and power to control airspeed and altitude.

c. Failure to establish and maintain the specified airspeed. Not applying correct pitch and power settings as required. Airspeed usually too high.

d. Excessive variations in altitude, heading and bank:

i. Not dividing attention as necessary.

ii. Not cross-checking instruments and applying necessary control application.

e. Rough or uncoordinated control technique.

i. Not recognizing and compensating for torque effect.

ii. Overcontrolling the aircraft.

f. Faulty trim technique. Not trimming aircraft as necessary.

g. Unintentional stall.

  i. Not recognizing imminent stall conditions.

  ii. Overcontrolling the aircraft resulting in a stall.

h. Inappropriate removal of hand from throttle. Not keeping hand on throttle during maneuver.

Basic Instrument
Maneuvers

7

A. General

1. Explain the value of learning basic maneuvers under simulated instrument conditions.

Developing the ability to maneuver an airplane for limited periods of time solely by reference to instruments and to follow radar instructions from ATC may become invaluable to a pilot when outside visual references are lost due to unexpected adverse weather. It must be emphasized to the student that this training does *not* prepare them for unrestricted operations in instrument weather conditions.

2. What are some common student errors in performance of the basic flight maneuvers? (FAA-S-8081-6AS)

Students commonly commit errors of:

a. "Fixation," "omission," and "emphasis" errors during instrument cross-check.

   i. Staring at one instrument and not cross-checking with other instruments.

   ii. Forgetting to include all instruments in scan.

   iii. Relying too heavily on one instrument for control of aircraft.

b. Improper instrument interpretation.

   i. Not interpreting information provided correctly.

   ii. Not sure which instruments provide pitch, bank, and power information.

c. Improper control applications. If instrument interpretation is incorrect, then control application will be incorrect also.

d. Failure to establish proper pitch, bank, or power adjustments during altitude, heading, or airspeed corrections.

   i. Not sure what instruments to look at when changing altitude, heading, or airspeed.

   ii. Not using the cross-check, interpret, and control method of attitude instrument flying.

Continued

    e. Faulty trim technique. Not using trim; using trim as a primary flight control.

    f. In performing straight, constant airspeed climbs, improper entry or level-off technique:

        i. Not using a specific pitch altitude and power setting for climb;

        ii. Not planning level-off and under or overshooting altitude.

    g. In performing turns to headings, improper entry or rollout technique:

        i. Not rolling into the turn gradually;

        ii. Not adjusting the pitch altitude to compensate for the loss of vertical lift in turn;

        iii. Not planning the rollout.

B. Straight-and-Level Flight

1. What is involved in straight-and-level flight by reference to instruments only?

Straight-and-level flight by reference to instruments under simulated instrument conditions involves utilizing the basic flight instruments to maintain a constant heading and altitude.

This is accomplished by:

a. Establishing a definite altitude.

b. Establishing a definite heading.

c. Establishing a cruise power setting and airspeed.

d. Trimming aircraft for "hands off" flight.

e. Maintaining level flight by positioning the miniature aircraft in relation to the horizon bar on the attitude indicator. Cross-check with the altimeter and vertical speed indicator.

f. Maintaining straight flight by referencing the directional gyro occasionally for any indication of heading change. Cross-check with the attitude indicator and turn coordinator.

g. Cross-checking the instruments, interpreting what they are telling you, and controlling the aircraft.

2. What are the standards expected of a student for straight-and-level flight under simulated instrument conditions? (FAA-S-8081-14S)

The student:

a. Shows knowledge of attitude instrument flying during straight-and-level flight.

b. Maintains straight-and-level flight solely by reference to instruments using proper instrument cross-check and interpretation, coordinated control application.

c. Maintains altitude, ±200 feet (60 meters); heading ±20 degrees; airspeed ±10 knots.

C. Straight, Constant Airspeed Climbs

1. Under simulated instrument conditions, what does a straight, constant airspeed climb involve?

A straight, constant airspeed climb involves a climb in which a constant heading is maintained at a predetermined airspeed. To accomplish this, the student should:

a. Raise the miniature aircraft to the approximate nose high indication appropriate to the predetermined climb speed.

b. Apply light back elevator pressure to initiate and maintain the climb attitude. The pressures will vary as the airplane decelerates.

c. Power may be advanced to the climb power setting simultaneously with the pitch change, or after the pitch change is established and the airspeed approaches climb speed.

d. Once the airplane stabilizes at a constant airspeed and attitude, the airspeed is the primary instrument for pitch and the heading indicator is primary for bank.

2. What are the standards expected of a student for the performance of a straight, constant airspeed climb by reference to instruments only? (FAA-S-8081-14S)

The student:

a. Shows knowledge of attitude instrument flying during straight, constant airspeed climbs.

b. Establishes climb configuration specified by examiner.

c. Transitions to climb pitch attitude and power setting on assigned heading using proper instrument cross-check and interpretation, and coordinated control application.

d. Demonstrates climbs solely by reference to instruments at a constant airspeed to specific altitudes in straight flight.

e. Levels off at the assigned altitude and maintains that altitude, ±200 feet (60 meters); heading ±20 degrees; airspeed ±10 knots.

D. Straight, Constant Airspeed Descents

1. Under simulated instrument conditions, what is involved in a straight, constant airspeed descent?

A straight constant airspeed descent involves a descent in which a constant heading is maintained at a predetermined airspeed. To accomplish this, the student should:

a. Reduce power to a predetermined value.

b. Maintain back pressure until the airspeed slows to the desired descent airspeed.

c. Lower the miniature aircraft to the approximate nose-low indication appropriate to the predetermined airspeed/rate of descent.

d. Once the airplane stabilizes at a constant airspeed/rate of descent and attitude, retrim the aircraft.

e. Control airspeed with pitch and rate of descent with power.

2. **What are the standards expected of a student for straight, constant airspeed descent executed under simulated instrument conditions?** (FAA-S-8081-14S)

The student:

a. Shows knowledge of attitude instrument flying during straight, constant airspeed descents.

b. Establishes descent configuration specified by examiner.

c. Transitions to descent pitch attitude and power setting on assigned heading using proper instrument cross-check and interpretation, and coordinator control application.

d. Demonstrates descents solely by reference to instruments at a constant airspeed to specific altitudes in straight flight.

e. Levels off at the assigned altitude and maintains that altitude, ±200 feet (60 meters); heading ±20 degrees; airspeed ±10 knots.

E. Turns to Headings

1. **What is involved in performing turns to headings by reference to instruments only?**

Performing turns to headings solely by reference to instruments involves changing or returning the aircraft heading to a desired heading. The process is as follows:

a. Apply coordinated aileron and rudder pressure in the desired direction of turn.

b. On the roll-in, use the attitude indicator to establish the approximate angle of bank, then check the turn coordinator for a standard rate turn indication.

c. During the roll-in, check the altimeter, vertical speed indicator, and attitude indicator for pitch adjustments necessary as the vertical lift component decreases with an increase in bank.

d. When rolling out, apply coordinated aileron and rudder pressures opposite the direction of turn.

Continued

   e. Lead the rollout by approximately 1/2 bank angle prior to the desired heading.

   f. Anticipate a possible gain in altitude on rollout and adjust pitch attitude as appropriate.

2. What are the standards expected of a student for performing turns to headings by reference to instruments only? (FAA-S-8081-14S)

The student:

a. Shows knowledge of attitude instrument flying during turns to headings.

b. Transitions to level-turn attitude using proper instrument cross-check and interpretation, and coordinated control application.

c. Demonstrates turns to headings solely by reference to instruments; maintains altitude, ±200 feet (60 meters); maintains a standard rate turn and rolls out on assigned heading, ±20 degrees; maintains airspeed ±10 knots.

F. Recovery From Unusual Flight Attitudes·

1. What are "unusual flight attitudes"?

When outside visual references are inadequate or lost, the VFR pilot is apt to unintentionally let the airplane enter a critical attitude (sometimes called an "unusual attitude"). In general, this involves an excessively nose-high attitude in which the airplane may be approaching a stall, or an extremely steep bank which may result in a steep downward spiral.

2. Discuss the importance of instrument training for recovery from an unusual attitude.

Since unusual attitudes are not intentional, they are often unexpected, and the reaction of an inexperienced or inadequately-trained pilot is usually instinctive rather than intelligent and deliberate. However, with practice the techniques for rapid and safe recovery from these critical attitudes can be learned.

3. Describe some techniques for recovery from unusual flight attitudes by reference to instruments.

Nose-high decreasing airspeed

a. Indicated by:

    i. Decreasing airspeed on airspeed indicator.

    ii. Increasing altitude on altimeter and positive rate on vertical speed indicator.

    ii. Bank on attitude indicator/heading indicator/turn coordinator.

b. Recover by:

    i. Reducing pitch attitude.

    ii. Simultaneously increasing power.

    iii. Leveling the wings as necessary.

Nose-low increasing airspeed

a. Indicated by:

    i. Increasing airspeed on airspeed indicator.

    ii. Decreasing altitude on altimeter and negative rate on vertical speed indicator.

    iii. Bank on attitude indicator/heading indicator/turn coordinator.

b. Recover by:

    i. Reducing power.

    ii. Leveling the wings.

    iii. Raising the nose gradually.

Note: The attitude indicator is normally the most useful instrument in determining the aircraft's attitude. However, extreme unusual attitudes may cause the gyros in the attitude indicator and heading indicator to tumble. In this situation the pitch attitude may be determined through combined use of the information provided by the airspeed indicator, altimeter, and vertical speed indicator. The turn coordinator may be used as the primary instrument in detecting bank angle.

4. What are the standards expected of a student for recovery from unusual flight attitudes by reference to instruments only? (FAA-S-8081-14S)

The student:

a. Shows knowledge of attitude instrument flying during unusual attitudes.

b. Recognizes unusual flight attitudes solely by reference to instruments; recovers promptly to a stabilized level flight attitude using proper instrument cross-check and interpretation and smooth, coordinated control application in correct sequence.

5. In simulated instrument conditions, what are some common student errors in recovering from unusual attitudes? (FAA-S-8081-6AS)

Students commonly commit errors of:

a. Failure to recognize an unusual flight attitude.

   i. Not interpreting the information provided correctly.

   ii. Not sure of which instruments provide pitch, bank, and power information.

b. Consequences of attempting to recover from an unusual flight attitude by feel rather than by instrument indications; not trusting instrument indications can lead to rapid altitude loss as a result of inadvertent steep spirals or power-on/power-off stalls.

c. Inappropriate control applications during recovery:

   i. Not following the correct sequence of actions in recovery;

   ii. In descents: increasing back pressure before leveling the wings, not reducing power in descent; overcompensating on recovery by increasing pitch attitude past that for level flight.

   iii. In climbs: not lowering the nose fast enough; lowering the nose too far; not adding power.

d. Failure to recognize, from instrument indications, when the airplane is passing through a level flight attitude:

   i. Not cross-checking instruments.

   ii. Not utilizing the attitude indicator and/or airspeed indicator/ altimeter to determine level flight.

Performance
Maneuvers

8

A. Steep Turns

1. Describe a steep turn.

Steep turns are those resulting from a degree of bank (more than approximately 45°) at which the overbanking tendency of an airplane overcomes stability, and the bank tends to increase unless pressure is applied to the aileron controls to prevent it. Maximum turning performance is attained and relatively high load factors are imposed.

2. What is the objective of practicing a steep turn?

The objective of the steep turn maneuver is to develop smoothness, coordination, orientation, division of attention, and control techniques while executing a high-performance turn.

3. Describe how to perform a steep turn maneuver.

a. Execute two clearing turns.

b. Establish a specific heading (and an outside reference) and altitude (at least 1,500 feet).

c. Roll into a 45° bank while maintaining altitude with necessary back pressure.

d. Complete a left and right 360° heading change. (It may be necessary to add additional power to maintain airspeed above a stall.)

e. Lead rollout to heading by one half the bank angle.

f. Use horizon and glare shield/cowling to maintain pitch attitude.

g. If loss of altitude occurs, decrease bank, add back pressure to regain altitude, and establish higher pitch attitude.

h. If gain of altitude occurs, increase bank (maximum 55°), decrease pitch attitude and or power.

4. What are the standards expected of a student for steep turns? (FAA-S-8081-12)

The student:

a. Shows knowledge of steep turns.

b. Selects an altitude that allows the task to be completed no lower than 1,500 feet AGL (460 meters) or the manufacturer's recommended altitude, whichever is higher.

c. Establishes manufacturer's recommended airspeed, or in its absence, examiner may designate a safe airspeed not to exceed V_A ±5 knots.

d. Enters a smooth, coordinated 360° steep turn with a 50° bank, ±5°, immediately followed by a 360° turn in the opposite direction.*

e. Divides attention between airplane control and orientation.

f. Rolls out on entry heading ±10°.

g. Maintains entry altitude throughout maneuver, ±100 feet (30 meters), and airspeed ±10 knots.

* *Private standards: Rolls into a coordinated 360° turn, maintains 45° bank, ±5°.*

5. What are some common student errors in performing steep turns? (FAA-S-8081-6AS)

a. Improper pitch, bank, and power coordination during entry and rollout.

   i. Increasing pitch attitude before establishing bank angle with resulting gain in altitude.

   ii. Not releasing back pressure in recovery with resulting gain in altitude.

   iii. Not enough or too much bank.

   iv. No use of power to assist in maintaining altitude.

b. Uncoordinated use of flight controls. Slipping or skidding through maneuver; a skid is common in turns to the right.

c. Inappropriate control applications—not making the necessary minor adjustments in altitude and bank throughout maneuver resulting in the need for major adjustments.

d. Improper technique in correcting altitude deviations—not reducing bank angle first before increasing back pressure when trying to correct for loss of altitude;

e. Loss of orientation—the initial heading from which the maneuver began is forgotten and/or the reference point outside can no longer be found.

f. Excessive deviation from desired heading during rollout.

   i. Not planning the rollout.

   ii. Not leading the heading to be rolled-out on by half the amount of bank in degree (i.e., 50° bank would require 25° of lead).

B. Chandelles

1. What is a chandelle?

A "chandelle" is a maximum performance climbing turn beginning from approximately straight-and-level flight, and ending at the completion of 180° of turn in a wings-level, nose-high attitude at the minimum controllable airspeed.

2. What is the objective of the chandelle maneuver?

The objective of performing chandelles is to develop the pilot's coordination, orientation, planning, and feel for maximum-performance flight, and to develop positive control techniques at varying airspeeds and attitudes. The maneuver demands that the maximum flight performance of the airplane be obtained. The airplane should gain the most amount of altitude possible for a given degree of bank and power setting without stalling.

3. Describe how to perform a chandelle.

a. Pick a prominent reference point and plan to perform the maneuver into the wind to prevent drift from the training area.

b. Perform a pre-maneuver checklist (GUMPS).

c. Establish maneuvering speed or the manufacturer's recommended speed.

Continued

d. Establish a coordinated 30° bank turn and plan on maintaining this bank angle for the first 90° of turn.

e. Apply takeoff or climb power and begin a climbing turn by smoothly applying back pressure to increase pitch attitude at a constant rate and to attain the highest pitch attitude as 90° of turn is completed.

f. When the turn has progressed 90° from the original heading, begin rolling out of the bank at a constant rate while maintaining a constant pitch attitude.

g. As the wings are being leveled at the completion of 180° of turn, the pitch attitude should be held momentarily while the airplane is at minimum controllable speed.

Note: *First 90°— constant bank, changing pitch*
Second 90°— constant pitch, changing bank

4. What are the standards expected of a student for performing chandelles? (FAA-S-8081-12)

The student:

a. Shows knowledge of performance factors associated with chandelles.

b. Selects an altitude that will allow the maneuver to be performed no lower than 1,500 feet AGL (460 meters) or manufacturer's recommended altitude, whichever is higher.

c. Establishes entry configuration at an airspeed no greater than the maximum entry speed recommended by the manufacturer (not to exceed V_A).

d. Establishes approximately, but does not exceed, 30° of bank.

e. Simultaneously applies specified power and pitch to maintain a smooth, coordinated climbing turn with constant bank to the 90°-point.

f. Begins a coordinated constant rate of rollout from the 90°-point to the 180°-point maintaining specified power and a constant pitch attitude that will result in a rollout within ±10° of desired heading and airspeed within +5 knots of power-on stall speed.

g. Reduces pitch attitude to resume straight-and-level flight at the final altitude attained, ±50 feet (20 meters).

5. What are some common student errors in the performance of chandelles? (FAA-S-8081-6AS)

a. Improper pitch, bank, and power coordination during entry and rollout.

   i. Initial pitch-up attitude to quick resulting in the aircraft approaching stall speed before 180° of direction change has occurred.

   ii. Initial pitch-up attitude too slow resulting in completion of maneuver at an airspeed well above the stall speed.

   iii. Initial bank angle exceeded resulting in completion of maneuver at an airspeed well above the stall speed.

   iv. Initial bank angle to shallow resulting in the aircraft approaching stall speed before 180° of direction change has occurred.

   v. Failure to add full power at the start of maneuver.

b. Uncoordinated use of flight controls — not using enough rudder to compensate for torque effect.

c. Improper planning and timing of pitch and bank attitude changes.

d. Factors related to failure in achieving maximum performance.

   i. Not maintaining appropriate bank angles during the maneuver.

   ii. Not maintaining appropriate pitch attitudes during the maneuver.

   iii. Not planning ahead.

e. A stall during the maneuver.

   i. Usually occurs in the last 90° of turn due to poor planning.

   ii. At the completion of the maneuver the airspeed should be approximately plus 5 knots above the stall speed.

f. Excessive deviation from desired heading during completion; allowing the heading to drift due to lack of torque correction.

C. Lazy Eights

1. What is a lazy eight?

A lazy eight consists of two 180° turns, in opposite directions, while making a climb and a descent in a symmetrical pattern during each of the turns.

2. Discuss the use and advantages of practicing lazy eights.

The objective of the lazy eight is to develop the pilot's feel for varying control forces, and the ability to plan and remain oriented while maneuvering the airplane with positive, accurate control. It requires constantly changing control pressures necessitated by changing combinations of climbing and descending turns at varying airspeeds. This is a maneuver often used to develop and demonstrate the pilot's mastery of the airplane in maximum performance flight situations.

3. Describe the steps involved in a lazy eight maneuver.

a. Pick a prominent reference point and plan to perform the maneuver into the wind to prevent drift from training area.

b. Perform clearing turns.

c. Perform pre-maneuver checklist (GUMPS).

d. Establish maneuvering speed or manufacturer's recommended speed.

e. Start the maneuver from level flight with a gradual climbing turn toward reference points.

f. At 45°-point: maximum pitch up attitude, bank 15° (approximately).

g. At 90°-point: bank approximately 30°, minimum airspeed, maximum altitude; level pitch attitude.

h. At 135°-point: maximum pitch down, bank 15° (approximately).

i. At 180°-point: level flight, entry airspeed, altitude same as entry altitude.

4. What are the standards expected of a student for performing lazy eights? (FAA-S-8081-12A)

The student:

a. Shows knowledge of performance factors associated with lazy eights.

b. Selects an altitude that will allow the task to be performed no lower than 1,500 feet AGL (460 meters) or the manufacturer's recommended altitude, whichever is higher.

c. Selects a prominent 90° reference point in the distance.

d. Establishes the recommended entry power and airspeed.

e. Plans and remains oriented while maneuvering airplane with positive, accurate control, and demonstrates mastery of the airplane.

f. Achieves the following throughout the task—

   i. constant change of pitch, bank, and turn rate.

   ii. altitude and airspeed consistent at the 90°-points, ±100 feet (30 meters) and ±10 knots respectively.

   iii. through proper power setting, attains the starting altitude and airspeed at the completion of the maneuver, ±100 feet (30 meters) and ±10 knots respectively.

   iv. heading tolerance ±10° at each 180°-point.

g. Continues task through at least two 180° circuits and resumes straight-and-level flight.

5. What are some common student errors in the performance of lazy eights? (FAA-S-8081-6AS)

a. Poor selection of reference points:

   i. Not selecting a reference point that is readily seen.

   ii. Picking a reference point that is too close.

b. Uncoordinated use of flight controls:

   i. Not correcting for torque effect.

   ii. Due to decreasing airspeed, considerable right rudder pressure must be gradually applied to counteract torque at the top of the eight in both right and left turns.

Continued

      iii. More right rudder pressure will be required during the climbing turn to the right than the turn to the left because more torque correction is needed to prevent yaw from decreasing the rate of turn.

      iv. In the left climbing turn the torque will tend to contribute to the turn consequently less rudder pressure is needed.

  c. Non-symmetrical loops resulting from poorly planned pitch and bank attitude changes:

      i. Initial pitch attitude too quick in the climbing turn resulting in the airplane stalling before reaching the 90°-point.

      ii. Initial pitch attitude too slow in the climbing turn resulting in the airplane reaching the 90°-point without attaining the minimum airspeed.

      iii. Initial bank angle excessive resulting in the airplane reaching the 90°-point before minimum airspeed is reached.

      iv. Allowing the pitch attitude in the last 90° of turn to become excessively low resulting in exceeding the initial entry airspeed and/or altitude.

  d. Inconsistent airspeed and altitude at key points—not planning ahead.

  e. Loss of orientation:

      i. Poorly selected reference point not readily seen.

      ii. Student not looking outside enough; poor division of attention.

  f. Excessive deviation from reference points—not planning ahead.

Ground
Reference
Maneuvers

9

A. Rectangular Course

1. Describe a rectangular course maneuver.

The rectangular course is a practice maneuver in which the ground track of the airplane is equidistant from all sides of a selected rectangular area on the ground. While performing the maneuver, the altitude and airspeed are held constant.

2. What are the advantages of practicing a rectangular course?

Like those of other ground track maneuvers, one of the objectives is to develop division of attention between the flightpath and ground references, while controlling the airplane at low altitudes and watching for other aircraft in the vicinity. Another objective is to develop recognition of drift toward or away from a line parallel to the intended ground track. This will be helpful in recognizing drift toward or from an airport runway during the various legs of the airport traffic pattern.

3. Describe the process of executing a rectangular course.

a. Select a square or rectangular field, the sides which are approximately 1 mile in length.

b. Establish an altitude 600 to 1,000 feet AGL (500 feet above obstructions).

c. Perform clearing turns.

d. Enter downwind and maneuver as necessary to maintain a ground track parallel to and a uniform distance from (approximately one fourth of a mile) field boundaries.

e. The bank in each of the turns should be adjusted to compensate for wind drift.

Note:

Upwind to crosswind: Shallow bank, increasing to medium, and less than 90° of turn.

Crosswind to downwind: Increasing to steepest bank necessary and more than 90° of turn.

Continued

Downwind to crosswind: Steepest bank necessary, decreasing to medium, and more than 90° of turn.

Crosswind to upwind: Medium bank, decreasing to shallow, and less than a 90° turn.

4. What are the standards expected of a student for a rectangular course? (FAA-S-8081-14S)

The student:

a. Shows knowledge of a rectangular course.

b. Determines wind direction and speed.

c. Selects ground reference area with an emergency landing area within gliding distance.

d. Plans maneuver to enter at traffic pattern altitude, at an appropriate distance from the selected reference area, 45° to the downwind leg, with the first circuit to the left.

e. Applies adequate wind drift correction during straight and turning flight to maintain a constant ground track around the rectangular surface area.

f. Divides attention between airplane control and ground track, maintains coordinated flight.

g. Exits at point of entry at same altitude and airspeed at which maneuver was started; reverses course as directed by examiner.

f. Maintains altitude, ±100 feet (30 meters); maintains airspeed, ±10 knots.

5. What are some common student errors in performing a rectangular course? (FAA-S-8081-6AS)

a. Poor planning, orientation, or division of attention:

   i. Not planning ahead.

   ii. Fixating on the field and forgetting to look for other air traffic.

   iii. Not selecting a proper distance from the field boundary.

    b. Uncoordinated flight control application:

        i. Turns are uncoordinated due to preoccupation with maneuver.

        ii. Not dividing attention correctly.

    c. Improper correction for wind drift:

        i. Failure to recognize or not correcting for wind drift.

        ii. No crab on straight-and-level segments, which results in drifting toward or away from field.

        iii. Not utilizing correct bank angles in turns.

        iv. Not turning the airplane when abeam the corners; instead the student turns the airplane when the wingtip is abeam the corner.

    d. Failure to maintain selected altitude or airspeed. Loss or gain in altitude due to lack of division of attention.

    e. Selection of a ground reference where there is no suitable emergency landing area within gliding distance. Poor planning when selecting area for maneuver.

B. S-Turns

1. Describe an "S-turn."

An S-turn is a practice maneuver in which the airplane's ground track describes semi-circles of equal radii on each side of a selected straight line on the ground. The straight line may be a road, fence, railroad, or section line which lies perpendicular to the wind, and should be of sufficient length for making a series of turns. The maneuver consists of crossing the road at a 90° angle and immediately beginning a series of 180° turns of uniform radius in opposite directions, recrossing the road at a 90° angle just as 180° of turn is completed. A constant altitude should be maintained throughout the maneuver. The altitude should be low enough to easily recognize drift but in no case lower than 500 feet above the highest obstruction.

2. What is the S-turn used for?

The objectives of this maneuver are, to develop the ability to compensate for wind drift during turns, orient the flightpath with ground references, and divide the pilot's attention while controlling the airplane at a relatively low altitude.

3. How are S-turns executed?

a. Select a straight ground reference line or road that lies 90° to the direction of the wind.

b. Establish an altitude 600 to 1,000 feet AGL (500 feet above obstructions).

c. Perform clearing turns.

d. Approach the road from the upwind side on a downwind heading and when directly over the road, roll into the first turn which should be the steepest.

e. As the airplane gradually turns back into the wind, begin to shallow the bank angle so as to arrive over the road as the wings roll level.

f. When directly over the road roll into the next turn, utilizing a shallow bank initially (upwind), then gradually increasing bank to the steepest (downwind).

4. What are the standards expected of a student for S-turns? (FAA-S-8081-14S)

The student:

a. Shows knowledge of S-turns.

b. Determines wind direction and speed.

c. Selects reference line with an emergency landing area within gliding distance.

d. Plans maneuver to enter at 600 to 1,000 feet (180 to 300 meters) AGL, perpendicular to the selected reference line, downwind, with the first series of turns to the left.

e. Applies adequate wind-drift correction to track a constant radius half-circle on each side of the selected reference line.

f. Divides attention between airplane control and ground track, maintains coordinated flight.

g. Reverses course as directed by examiner, and exits at point of entry at same altitude and airspeed at which maneuver was started.

5. What are some common student errors in performing S-turns? (FAA-S-8081-6AS)

a. Faulty entry technique:

  i. Entering upwind.

  ii. Not clearing the area.

  iii. Entering at an improper altitude.

b. Poor planning, orientation, or division of attention:

  i. Appropriate bank angles utilized too soon or too late due to lack of planning.

  ii. Forgets wind direction in relation to location.

  iii. Not dividing attention inside and outside equally.

c. Uncoordinated flight control application. Not dividing attention inside and outside resulting in uncoordinated turns.

d. Improper correction for wind drift:

  i. Forgetting where wind is in relation to the airplane;

  ii. Not sure when to use steep and shallow bank angles.

e. A non-symmetrical ground track: not compensating for the wind, especially on the upwind side.

f. Failure to maintain selected altitude or airspeed. Not dividing attention inside and outside resulting in a loss or gain in altitude.

g. Selection of a ground reference line where there is no suitable emergency landing area within gliding distance — not planning ahead.

C. Turns Around a Point

1. What are "turns around a point"?

This training maneuver involves flying the airplane in two or more circles of uniform radii or distance from a prominent ground reference point using a maximum bank of approximately 45° while maintaining a constant altitude.

2. What is objective of practicing turns around a point?

The objective, as with other ground reference maneuvers, is to help the pilot develop the ability to subconsciously control the airplane while at relatively low altitudes, dividing attention between the flight path and ground references, and watching for other air traffic in the vicinity.

3. How are turns around a point executed?

a. Select a prominent point on the ground.

b. Establish an altitude 600 to 1,000 feet AGL (500 feet above obstructions).

c. Perform clearing turns.

d. Enter the maneuver downwind at a distance equal to the desired radius of turn.

e. Roll into a bank when abeam reference. If any significant wind is present this will be the steepest bank in the maneuver (highest ground speed).

f. Decrease the bank angle as the aircraft turns into the wind (low ground speed).

Note:

Downwind turn: Steepest bank.
Downwind to crosswind: Gradually decrease to medium bank.
Crosswind to upwind: Gradually decrease from medium to shallow bank.
Upwind to crosswind: Gradually increase from shallow to medium bank.
Crosswind to downwind: Increase from medium to steep bank.

4. What are the standards expected of a student for turns around a point? (FAA-S-8081-14S)

The student:

a. Shows knowledge of turns around a point.

b. Determines wind direction and speed.

c. Selects reference point with an emergency landing area within gliding distance.

d. Plans maneuver to enter at 600 to 1,000 feet (180 to 300 meters) AGL, at an appropriate distance from the reference point, with the airplane headed downwind and the first turn to the left.

e. Applies adequate wind-drift correction to track a constant radius circle around the selected reference point with a bank of approximately 45° at the steepest point in the turn.

f. Divides attention between airplane control and ground track, maintains coordinated flight.

g. Completes two turns, exits at point of entry at same altitude and airspeed at which maneuver was started, and reverses course as directed by examiner.

h. Maintains altitude, ±100 feet (30 meters); maintains airspeed, ±10 knots.

5. What are some common student errors in performing turns around a point? (FAA-S-8081-6AS)

a. Faulty entry technique:

   i. Entering upwind.

   ii. Not clearing the area.

  iii. Enters at improper altitude.

  iv. Not utilizing the steepest bank on the initial turn from downwind.

Continued

b. Poor planning, orientation, or division of attention:

   i. Appropriate bank angles utilized too soon or too late due to lack of planning.

   ii. Forgets wind direction in relation to location.

   iii. Not dividing attention inside and outside equally.

c. Uncoordinated flight control application. Not dividing attention inside and outside resulting in uncoordinated turns.

d. Improper correction for wind drift:

   i. Forgetting where the wind is in relation to the airplane.

   ii. Not sure when to use steep and shallow bank angles.

e. Failure to maintain selected altitude or airspeed. Not dividing attention inside and outside resulting in a loss or gain in altitude.

f. Selection of a ground reference point where there is no suitable emergency landing area within gliding distance—not planning ahead.

D. Eights-On-Pylons

1. Describe "eights-on-pylons."

Eights-on-pylons is a training maneuver which involves flying the airplane in circular paths, alternately left and right, in the form of a figure-8 around two selected points or pylons on the ground. No attempt is made to maintain a uniform distance from the pylon. Instead, the airplane is flown at such an altitude and airspeed that a line parallel to the airplane's lateral axis, and extending from the pilot's eye, appears to pivot on each of the pylons.

2. What is the objective of the eights-on-pylons maneuver?

While not truly a ground track maneuver, the objective is similar: to develop the ability to maneuver the airplane accurately while dividing one's attention between the flight path and maintaining a pivotal position on selected pylons on the ground.

3. How are eights-on-pylons executed?

a. Select two points on the ground along a line which lies 90° to the wind. The points should be prominent and should be adequately spaced to provide planning for the turns (approximately 3 to 5 seconds straight-and-level flight).

b. Establish the pivotal altitude.

c. Perform clearing turns.

d. Begin the maneuver by flying diagonally crosswind between the pylons to a point downwind from the first pylon so that the first turn can be made into the wind.

e. As the airplane approaches the pylon a turn should be started by lowering the wing to place the pilot's line of sight reference on the pylon.

f. As the airplane heads into the wind the groundspeed decreases, and consequently, the pivotal altitude is lower and the airplane must descend to hold the reference line on the pylon.

g. As the turn progresses on the upwind side of the pylon, the wind becomes more of a crosswind and drifts the airplane closer to the pylon. Since a constant distance is not required no correction should be applied.

h. With the airplane drifting closer to the pylon, the angle of bank must be increased to hold the reference line on the pylon.

i. If the reference line appears to move ahead of the pylon the pilot should increase altitude.

j. As the airplane turns toward a downwind heading, the rollout from the turn should be started to allow the airplane to proceed diagonally to a point on the downwind side of the second pylon.

k. The rollout must be completed in the proper crab angle to correct for wind drift so that the airplane will arrive at a point downwind from the second pylon the same distance it was from the first pylon.

l. Upon reaching second pylon a turn is started in the opposite direction by lowering the upwind wing to again place the pilot's line of sight reference on the pylon.

4. What are the standards expected of a student for performing eights-on-pylons? (FAA-S-8081-12A)

The student:

a. Shows knowledge of eights-on-pylons including the relationship of groundspeed change to performance of the maneuver.

b. Determines approximate pivotal altitude.

c. Selects suitable pylons, considering emergency landing areas, that will permit approximately 3 to 5 seconds of straight-and-level flight between them.

d. Attains proper configuration and airspeed prior to entry.

e. Applies necessary corrections so that line-of-sight reference line remains on the pylon with minimum longitudinal movement.

f. Exhibits proper orientation, division of attention, and planning.

g. Applies necessary wind-drift correction to track properly between pylons.

h. Holds pylon using appropriate pivotal altitude avoiding slips and skids.

5. What are some common student errors in performing eights-on-pylons? (FAA-S-8081-6AS)

a. Faulty entry technique:

   i. Poor pylon selection.

   ii. Not entering by flying diagonally crosswind between the pylons.

   iii. Not clearing the area; not entering at pivotal altitude.

b. Poor planning, orientation, and division of attention:

   i. Not planning for changes in groundspeed.

   ii. Losing the pylon.

   iii. Forgetting wind direction in relation to location.

   iv. Not dividing attention inside and outside equally.

c. Uncoordinated flight control application:

    i. Attempting to hold a pylon by use of rudder.

    ii. Not dividing attention inside and outside resulting in unco-ordinated turns.

d. Use of an improper line of sight reference; not using a line of sight reference parallel to the lateral axis of the airplane.

e. Application of rudder alone to maintain the line of sight on pylon:

    i. When reference line moves ahead of pylon, the pilot will tend to press the inside rudder to yaw the wing backward.

    ii. When the reference line moves behind the pylon, the pilot will press the outside rudder to yaw the wing forward.

f. Improper timing of turn entries and rollouts:

    i. Rolling in too soon or rolling out too late.

    ii. Rolling in should be started just before the reference line reaches pylon.

    iii. The rollout should be started to allow the airplane to proceed diagonally to a point on the downwind side of the second pylon.

g. Improper correction for wind drift between pylons—not compensating for wind drift when between pylons. The distance from second pylon will not be equal to distance from first pylon.

h. Selection of pylons where there is no suitable emergency landing area within gliding distance — not planning ahead.

Approaches and Landings

10

A. Normal or Crosswind Approach and Landing

1. Describe a normal approach and landing.

A normal approach and landing involves the techniques for what is considered a "normal" situation; that is, when engine power is available, the wind is light and the final approach is made directly into the wind, the final approach path has no obstacles, and the landing surface is firm and of ample length to gradually bring the airplane to a stop.

2. What is the learning objective of a normal approach and landing?

The objective of a good stabilized approach and landing is to establish an angle of descent and airspeed that will permit the airplane to reach the desired touchdown point at an airspeed which will result in a minimum floating just before touchdown.

3. List the steps toward achieving a normal approach and landing.

a. Establish a normal traffic pattern.

b. At midfield, on downwind, complete a pre-landing checklist.

c. Abeam the touch down point on downwind, reduce power.

d. Maintain altitude and level pitch attitude momentarily to dissipate airspeed.

e. Retrim aircraft to establish airspeed within flap operating range (white arc).

f. Lower flaps to 10°.

g. Establish initial approach airspeed (1.4 x V_{SO}); retrim if necessary.

h. At an approximate 45°-point from the landing threshold, clear for traffic and turn base.

i. Extend flaps and retrim if necessary to maintain approach airspeed; apply wind drift correction.

Continued

    j. Lead turn to final to roll out on runway extended centerline.

    k. Once the field is assured extend final flaps.

    l. Adjust pitch for desired airspeed and power for rate of descent.

    m. As airplane approaches runway and landing is assured slowly close the throttle and start the landing flare about 5 to 10 feet above the runway.

    n. Continue applying back pressure as the airplane decelerates.

    o. Land on the main wheels first.

    p. Brake as necessary.

4. What is involved in a crosswind approach and landing?

A crosswind approach and landing involves techniques utilized when the wind is blowing across rather than parallel to the runway direction. Many runways or landing areas are such that landings must be made while the wind is blowing across rather than parallel to the landing direction; therefore, all pilots should be prepared to cope with these situations when they arise.

5. What is the procedure in a crosswind approach and landing?

    a. Utilize the same procedures used for a normal approach.

    b. On final, lower the upwind wing as necessary to control lateral drift over the extended runway centerline.

    c. Use opposite rudder to align the longitudinal axis of the airplane with the extended runway centerline.

    d. Maintain and adjust the control deflections as necessary to track the extended centerline.

    e. As the airspeed slows during the roundout and flare, control deflections must be increased as necessary to obtain the desired effect.

    f. After touchdown, slowly increase aileron deflection into the crosswind to assist in directional control.

6. What are the standards expected of a student for a normal or crosswind approach and landing? (FAA-S-8081-12A)

The student:

a. Shows knowledge of normal and crosswind approach and landing.

b. Considers wind conditions, landing surface, and obstructions.

c. Selects a suitable touchdown point.

d. Establishes recommended approach and landing configuration; adjusts power and attitude as required.

e. Maintains a stabilized approach and recommended airspeed with gust factor applied, ±5 knots.

Private PTS: Maintains a stabilized approach and the recommended approach speed, or in its absence, not more than 1.3 V_{SO}, ±10 knots, with gust factor applied.

f. Makes smooth, timely, and correct control application during roundout and touchdown.

g. Remains aware of the possibility of wind shear and/or wake turbulence.

h. Touches down smoothly at approximate stalling speed, at a specified point at or within 200 feet (60 meters) beyond a specified point with no drift, and with the airplane's longitudinal axis aligned with and over the runway centerline.

Private PTS: Touches down smoothly at the approximately stalling speed, at or within 400 feet (120 meters) beyond a specified point.

i. Maintains crosswind correction and directional control throughout approach and landing.

j. Completes appropriate checklist.

Note: If a crosswind condition does not exist, the applicant's knowledge of the crosswind elements shall be evaluated through oral testing.

7. What are some common student errors in performing normal or crosswind approach and landings?
(FAA-S-8081-6AS)

a. Improper use of landing performance data and limitations. Not calculating landing distances correctly.

b. Failure to establish the approach and landing configuration at the appropriate time or in the proper sequence. Not following established procedure.

c. Failure to establish and maintain a stabilized approach:

 i. Not establishing correct airspeeds for downwind, base and final segments.

 ii. Not reducing power as necessary to control rate of descent.

 iii. Not utilizing flaps as necessary to control speed and rate of descent.

 iv. Not correcting for wind drift on downwind, base, and final.

d. Inappropriate removal of hand from throttle. Not keeping hand on throttle for power increases and reductions.

e. Improper techniques during roundout and touchdown:

 i. Rounding out too late resulting in, if not corrected, a hard landing followed by a bounce and a stall and another hard landing.

 ii. Rounding out too high resulting in, if not corrected, an eventual loss in airspeed followed by a high sink rate and a hard landing.

 iii. Rounding out and "ballooning" down the runway; usually caused by misjudging the rate of descent and overcontrolling.

 iv. Rounding out and "floating" down the runway; usually caused by excessive airspeed on final approach.

 v. Touchdown followed by a bounce as a result of an excessive rate of descent on final.

 vi. Touchdown with the aircraft drifting sideways as a result of not applying enough "wing low" into the crosswind.

 vii. Touchdown in a crab as a result of allowing a wing to rise on roundout or in the flare.

f. Poor directional control after touchdown:

   i. Allowing a wing to rise after touchdown.

   ii. Allowing touchdown while in a crab.

   iii. Overcontrolling with rudder.

   iv. Ground loop: may be caused by a crosswind or overcontrolling airplane; this problem can be significant in tailwheel aircraft.

g. Improper use of brakes:

   i. Not utilizing aerodynamic braking.

   ii. Excessive use of brakes.

   iii. Skidding the tires.

B. Forward Slip to a Landing

1. What is a forward slip?

A forward slip is a descent with one wing lowered and the airplane's longitudinal axis at an angle to the flightpath.

2. What is the purpose of a forward slip to a landing?

The primary purpose of forward slips is to dissipate altitude without increasing the airplane's speed, particular in airplanes not equipped with flaps. There are many circumstances requiring the use of forward slips, such as in a landing approach over obstacles and in making forced landings, when it is always wise to allow an extra margin of altitude for safety in the original estimate of the approach.

3. How is a forward slip to a landing performed?

a. Reduce power to idle.

b. The wing on the side toward which the slip is to be made should be lowered by use of ailerons.

Continued

c. Simultaneously, the airplane's nose must be yawed in the oppo-site direction by applying opposite rudder so that the airplane's longitudinal axis is at an angle to its original flight path.

d. The degree to which the nose is yawed in the opposite direction from the bank should be such that the original ground track is maintained.

e. The pitch should be adjusted as necessary to maintain the appropriate airspeed.

f. Discontinuing the slip is accomplished by leveling the wings and simultaneously releasing rudder pressure while readjusting pitch attitude to the normal glide attitude.

4. What should a pilot be aware of regarding instrument error in a forward slip?

Because of the location of the pitot tube and static vents, airspeed indicators in some airplanes may have considerable error when the airplane is in a slip. The pilot must be aware of this possibility and recognize a properly performed slip by the attitude of the airplane, the sound of the airflow, and the feel of the flight controls. For-ward slips with the wing flaps extended should not be done in airplanes wherein the manufacturer's operating instructions pro-hibit such operation.

5. What are the standards expected of a student in performing a forward slip to a landing?
(FAA-S-8081-14S)

The student:

a. Shows knowledge of a forward slip to a landing.

b. Considers wind conditions, landing surface and obstructions, and selects the most suitable touchdown point.

c. Establishes slipping attitude at the point from which a landing can be made using recommended approach, landing configura-tion and airspeed; adjusts pitch attitude and power as required.

d. Maintains ground track aligned with runway centerline and airspeed which results in minimum float during roundout.

e. Makes smooth, timely, and correct control application during recovery from the slip, roundout, and touchdown.

f. Touches down smoothly at approximate stalling speed, at or within 400 feet (120 meters) beyond a specified point, with no side drift, and with the airplane's longitudinal axis aligned with and over runway centerline.

g. Maintains crosswind correction and directional control throughout approach and landing.

h. Completes appropriate checklist.

6. What are some common student errors in performing a forward slip to a landing? (FAA-S-8081-6AS)

a. Improper use of landing performance data and limitations:
   i. Not calculating landing distances correctly.
   ii. Performing a slip with flaps in an airplane in which this operation is prohibited.

b. Failure to establish approach and landing configuration at the appropriate time or in the proper sequence. Not following established procedure.

c. Failure to maintain a stabilized slip:
   i. Not reducing power.
   ii. Not applying sufficient aileron and rudder to establish forward slip.
   iii. Not maintaining equal amounts of aileron and rudder input.
   iv. Not maintaining a safe airspeed while in the slip.

d. Inappropriate removal of hand from throttle. Not keeping hand on throttle for power increases and reductions.

e. Improper technique during transition from the slip to the touchdown:
   i. The longitudinal axis of the airplane is not aligned with the runway centerline as a result of not allowing sufficient time between recovery from the slip and touch down.
   ii. Underestimating rate of sink during slip and landing hard.

Continued

    f. Poor directional control after touchdown:

        i. Allowing a wing to rise after touchdown.

        ii. Allowing touchdown while in a crab.

        iii. Overcontrolling with rudder.

        iv. Ground loop: may be caused by a crosswind or overcontrolling airplane; this problem can be significant in tailwheel aircraft.

    g. Improper use of brakes:

        i. Not utilizing aerodynamic braking.

        ii. Excessive use of brakes.

        iii. Skidding the tires.

C. Go-Around

1. Describe a go-around.

A go-around is a procedure utilized when it may be advisable for safety reasons to discontinue the landing approach and make another approach under more favorable conditions. A go-around involves procedures including establishing full power, drag reduction and traffic pattern procedures.

2. Why is it important for the pilot to be ready for the go-around?

The earlier a dangerous situation is recognized and the sooner the landing is rejected and the go-around started, the safer the procedure will be. Extremely low base to final turns, overshot or low final approaches, the unexpected appearance of hazards on the runway, wake turbulence from a preceding airplane, or overtaking another airplane on the approach are hazardous conditions that would demand a go-around.

3. How is the go-around executed?

    a. Make the decision to go-around as early as possible.

    b. Simultaneously increase pitch attitude to stop the descent and apply takeoff power.

c. If the aircraft has been trimmed for the approach, expect to hold considerable forward elevator pressure to maintain a safe climb attitude.

d. After the descent has been stopped, the landing flaps may be partially retracted or placed in the takeoff position, as recommended by the manufacturer.

e. Roughly retrim the aircraft to relieve control pressure.

f. Establish a best angle (V_X) or best rate (V_Y) of climb as appropriate.

g. Retract the landing gear when a positive rate of climb has been established.

h. Retrim aircraft.

4. What are the standards expected of a student for a go-around? (FAA-S-8081-12A)

The student:

a. Shows knowledge of a go-around.

b. Makes timely decision to discontinue approach to landing.

c. Applies maximum allowable power immediately and establishes pitch attitude that will stop the descent.*

d. Retracts flaps to approach setting.

e. Retracts landing gear after a positive rate-of-climb is established, (or as specified).

f. Trims airplane to accelerate to V_Y before final flap retraction, then climbs at V_Y, ±5 knots.

g. Maneuvers to the side of runway/landing area to clear and avoid (simulated) conflicting traffic.

h. Maintains maximum available power to a safe maneuvering altitude, then sets climb power.

i. Maintains proper wind-drift correction and obstruction clearance throughout transition to climb.

j. Completes appropriate checklist.

* *Private PTS: Applies takeoff power immediately and transitions to the climb pitch attitude for V_Y, +10/-5 knots.*

5. What are some common student errors in the performance of a go-around? (FAA-S-8081-6AS)

a. Failure to recognize a situation where a go-around is necessary. Not recognizing unsafe conditions that warrant a go-around.

b. Hazards of delaying a decision to go around. Not making a decision until the last minute can make the go around an unsafe procedure.

c. Improper power application:

   i. Not applying full power.

   ii. Failure to remove carburetor heat.

   iii. Failure to adjust propeller to low pitch high RPM first.

   iv. Power application not smooth.

d. Failure to control pitch attitude:

   i. On initial full power application, not controlling the sharp nose up attitude that will occur.

   ii. Not establishing the pitch attitude for best angle or best rate of climb.

e. Failure to compensate for torque effect: not correcting for torque effect in climb.

f. Improper trim technique: not initially retrimming aircraft to relieve heavy control pressure.

g. Failure to maintain recommended airspeeds; not establishing and maintaining best angle or best rate of climb as appropriate.

h. Improper wing flaps or landing gear retraction procedure:

   i. Not retracting flaps, retracting flaps all at once, or too soon considering altitude.

   ii. Retracting the landing gear before a positive rate of climb has been established.

D. Short-Field Approach and Landing

1. Describe a short-field approach and landing.

The short-field approach and landing is a maximum performance operation that requires the use of procedures and techniques for the approach and landing at fields which have a relatively short landing area or where an approach must be made over obstacles which limit the available landing area. It is one of the most critical of the maximum performance operations, since it requires that the pilot fly the airplane at one of its crucial performance capabilities while close to the ground in order to safely land within confined areas.

2. Why is it important to practice short-field approaches and landings?

To land within a short field or confined area, the pilot must have precise, positive control of the rate of descent and airspeed to produce an approach that will clear any obstacles, result in little or no floating during the roundout, and permit the airplane to be stopped in the shortest possible distance.

3. How is a short-field approach and landing performed?

a. Establish a normal traffic pattern.

b. At midfield, on downwind, complete a pre-landing checklist.

c. Abeam the touch down point on downwind, reduce power.

d. Maintain altitude and level pitch attitude momentarily to dissipate airspeed.

e. Retrim aircraft to establish airspeed within flap operating range (white arc).

f. Lower flaps to 10°.

g. Establish initial approach airspeed (1.4 x V_{SO}); Retrim if necessary.

h. At an approximate 45°-point from the landing threshold (30°-point for a short field with obstacle), clear for traffic and turn base.

Continued

   i. Extend flaps and retrim if necessary to maintain approach airspeed; apply wind drift correction.

   j. Lead turn to final to roll out on runway extended centerline.

  k. Extend full flaps on final.

   l. Adjust pitch for an approach airspeed of 1.3 x V_{SO} and adjust power to control rate of descent.

  m. If landing over a 50-foot obstacle, when clear, adjust pitch attitude slightly to establish rate of descent. Do not reduce power until in ground effect.

  n. If landing with no obstacle, adjust descent angle to land just inside of the desired touchdown point.

  o. After landing, identify and retract flaps, and apply maximum braking and full elevator back pressure.

4. What are the standards expected of a student for short-field approaches and landings? (FAA-S-8081-12A)

The student:

a. Shows knowledge of short-field approach and landing.

b. Considers wind conditions, landing surface, and obstructions.

c. Selects the most suitable touchdown point.

d. Establishes recommended approach and landing configuration adjusts attitude and power as required.

e. Maintains a stabilized approach, and recommended airspeed (or in its absence not more than 1.3 V_{SO}), with gust correction factor applied, ±5 knots.

Private PTS: Maintains a stabilized approach and the recommended approach airspeed, or in its absence not more than 1.3 V_{SO}, +10/-5 knots, with gust factor applied.

f. Makes smooth, timely, and correct control application during roundout and touchdown.

g. Remains aware of the possibility of wind shear and/or wake turbulence.

h. Touches down at a specified point at or within 100 feet (30 meters) beyond a specified point, with little or no float, with no drift, and with the airplane's longitudinal axis aligned with and over the runway centerline.

Private PTS: Touches down smoothly at the approximate stalling speed, at or within 200 feet (60 meters) beyond a specified point.

i. Maintains crosswind correction and directional control throughout the approach and landing.

j. Applies brakes, as necessary, to stop in the shortest distance consistent with safety.

k. Completes appropriate checklist.

5. What are some common student errors in the performance of short-field approaches and landings? (FAA-S-8081-6AS)

a. Improper use of landing performance data and limitations—not calculating landing distances correctly.

b. Failure to establish the approach and landing configuration at the appropriate time or in the proper sequence. Not following established procedure.

c. Failure to establish and maintain a stabilized approach:

 i. Not establishing the correct airspeeds for downwind, base and final segments.

 ii. Not reducing power as necessary to control rate of descent.

 iii. If short field over an obstacle:

 — when clear of obstacle, reducing power resulting in possible stall

 — diving for runway resulting in excessive airspeed and floating

 iv. Not utilizing flaps as necessary to control speed and rate of descent.

 v. Not correcting for wind drift on downwind, base, and final.

d. Improper technique in use of power, wing flaps, and trim:

 i. Not using pitch and power to control airspeed and rate of descent.

 ii. Not extending flaps as necessary.

 iii. Not trimming aircraft for appropriate airspeed.

Continued

e. Inappropriate removal of hand from throttle — not keeping hand on throttle for power increases and reductions.

f. Improper techniques during roundout and touchdown:

 i. Rounding out too late resulting in, if not corrected, a hard landing followed by a bounce and a stall and another hard landing.

 ii. Rounding out too high resulting in, if not corrected, an eventual loss in airspeed followed by a high sink rate and a hard landing.

 iii. Rounding out and "ballooning" down the runway; usually caused by misjudging the rate of descent and over-controlling.

 iv. Rounding out and "floating" down the runway; usually caused by excessive airspeed on final approach.

 v. Touchdown followed by a bounce as a result of an excessive rate of descent on final.

g. Poor directional control after touchdown:

 i. Allowing a wing to rise after touchdown.

 ii. Allowing touchdown while in a crab.

 iii. Overcontrolling with rudder.

 iv. Ground loop—may be caused by a crosswind or over-controlling airplane; this problem can be significant in tail-wheel aircraft.

h. Improper use of brakes:

 i. Not utilizing aerodynamic braking.

 ii. Excessive use of brakes.

 iii. Skidding the tires.

E. Soft-Field Approach and Landing

1. What is a soft-field approach and landing?

Landing on fields that are rough or have soft surfaces, such as snow, sand, mud, or tall grass requires unique techniques. When landing on such surfaces, the pilot must control the airplane in a manner that the wings support the weight of the airplane as long as practical, to minimize drag and stresses imposed on the landing gear by the rough or soft surface. The approach for the soft-field landing is similar to the normal approach used for operating into long firm landing areas. The major difference between the two is that during the soft field landing, the airplane is held 1 to 2 feet off the surface as long as possible to dissipate the forward speed sufficiently to allow the wheels to touch down gently at minimum speed.

2. What is the objective for practicing soft-field approaches and landings?

Operations into and out of airports with paved runways will not always be possible. Soft-field approach and landing techniques may be utilized when normal approach and landing procedures will not be effective or particularly safe.

3. How is a soft field approach and landing performed?

a. Establish a normal traffic pattern.

b. At midfield, on downwind, complete a pre-landing checklist.

c. Abeam the touch down point on downwind, reduce power.

d. Maintain altitude and level pitch attitude momentarily to dissipate airspeed.

e. Retrim aircraft to establish airspeed within flap operating range (white arc).

f. Lower flaps to 10°.

g. Establish initial approach airspeed (1.4 x V_{SO}); retrim if necessary.

h. At an approximate 45°-point from the landing threshold (30°-point for a short field with obstacle), clear for traffic and turn base.

i. Extend flaps and retrim if necessary to maintain approach airspeed; apply wind-drift correction.

j. Lead turn to final to roll out on runway extended centerline.

k. Extend full flaps on final.

l. Adjust pitch for an approach airspeed of 1.3 x V_{SO} and adjust power to control the rate of descent.

Continued

m. Touchdown at the lowest possible airspeed with the airplane in a nose-high pitch attitude.

n. After the main wheels touch the surface, hold sufficient back elevator pressure to keep the nose wheel off the ground until it can no longer aerodynamically be held off the field surface.

o. Gently lower the nose wheel to the surface.

p. A slight addition of power during and immediately after the touchdown will aid in easing the nose wheel down.

q. Avoid the use of brakes.

r. Increase power, as necessary, to keep the airplane moving and from becoming stuck in the surface.

4. What are the standards expected of a student for soft-field approaches and landings? (FAA-S-8081-12A)

The student:

a. Shows knowledge of a soft-field approach and landing.

b. Considers wind conditions, landing surface, and obstructions.

c. Selects the most suitable touchdown point.

d. Establishes recommended approach and landing configuration; and adjusts power and pitch attitude as required.

e. Maintains stabilized approach and recommended airspeed, or in its absence, not more than 1.3 V_{SO}, with gust factor applied, ±5 knots.*

f. Makes smooth, timely, and correct control application during roundout and touchdown.

g. Maintains crosswind correction and directional control throughout approach and landing.

h. Touches down softly, with no drift, and with the airplane's longitudinal axis aligned with the landing surface.

i. Maintains proper position of flight controls and sufficient speed to taxi on soft surface.

j. Completes appropriate checklists.

Private PTS: Maintains a stabilized approach and recommended approach airspeed, or in its absence not more than 1.3 V_{SO}, +10/-5 knots, with gust factor applied.

5. **What are some common student errors in the performance of soft-field approaches and landings?** (FAA-S-8081-6AS)

a. Improper use of landing performance data and limitations. Not calculating landing distances correctly.

b. Failure to establish approach and landing configuration at the appropriate time or in the proper sequence. Not following established procedure.

c. Failure to establish and maintain a stabilized approach:

   i. Not establishing correct airspeeds for downwind, base and final segments.

   ii. Not reducing power as necessary to control rate of descent.

   iii. Not utilizing flaps as necessary to control speed and rate of descent.

   iv. Not correcting for wind drift on downwind, base, and final.

d. Failure to consider the effect of wind and landing surface:

   i. Not inspecting area first to determine landing surface.

   ii. Not considering the effect of a headwind in slowing the airplane's forward speed on touchdown.

e. Improper technique in use of power, wing flaps or trim:

   i. Not using pitch and power to control airspeed and rate of descent.

   ii. Not extending flaps as necessary.

   iii. Not trimming aircraft for appropriate airspeed.

f. Inappropriate removal of hand from throttle. Not keeping hand on throttle for power increases and reductions.

g. Improper techniques during roundout and touchdown:

   i. Rounding out too late resulting in, if not corrected, a hard landing followed by a bounce and a stall and another hard landing.

   ii. Rounding out too high resulting in, if not corrected, an eventual loss in airspeed followed by a high sink rate and a hard landing.

Continued

      iii. Rounding out and "ballooning" down the runway; usually caused by misjudging the rate of descent and over-controlling.

      iv. Rounding out and "floating" down the runway; usually caused by excessive airspeed on final approach.

      v. Touchdown followed by a bounce as a result of an excessive rate of descent on final.

h. Failure to hold back elevator pressure after touchdown. Not maintaining sufficient back elevator pressure to keep weight off of the nose wheel and prevent a heavy load on the nose gear causing the nose wheel to dig in.

i. Closing the throttle to soon after touchdown:

      i. Not maintaining some power to increase elevator effectiveness so the weight may be kept off of the nose wheel as long as possible.

      ii. Not maintaining some power to keep aircraft moving and prevent it from becoming stuck.

j. Poor directional control after touchdown:

      i. Allowing a wing to rise after touchdown.

      ii. Allowing touchdown while in a crab.

      iii. Overcontrolling with rudder.

      iv. Ground loop—may be caused by a crosswind or over-controlling airplane; this problem can be significant in tail-wheel aircraft.

k. Improper use of brakes:

      i. Not utilizing aerodynamic braking.

      ii. Excessive use of brakes.

      iii. Skidding the tires.

U.S. Department
of Transportation
**Federal Aviation
Administration**

Advisory Circular
61-65D

Subject: **Certification: Pilots, and Flight and Ground Instructors**

Date: 9/20/99   Initiated By: AFS-800

1. **PURPOSE.** This advisory circular (AC) provides guidance for pilots, flight instructors, ground instructors, and examiners on the certification standards, knowledge test procedures, and other requirements contained in Title 14 of the Code of Federal Regulations (14 CFR) part 61.

Contents

2. **CANCELLATION.** AC 61-65C, Certification: Pilots and Flight Instructors, dated February 11, 1991, is canceled.

3. **RELATED READING MATERIAL.**

   a. AC 61-98, Currency and Additional Qualification Requirements for Certificated Pilots (current edition).

   b. AC 61-101, Presolo Written Test.

   c. AC 61-107, Operations of Aircraft at Altitudes Above 25,000 Feet MSL and/or MACH Numbers (Mmo) Greater Than .75.

   d. FAA-G-8082-1, Airline Transport Pilot, Aircraft Dispatcher, and Flight Navigator Knowledge Test Guide.

   e. FAA-G-8082-5, Commercial Pilot Knowledge Test Guide.

   f. FAA-G-8082-7, Flight and Ground Instructor Knowledge Test Guide.

   g. FAA-G-8082-13, Instrument Rating Knowledge Test Guide.

   h. FAA-G-8082-17, Recreational Pilot and Private Pilot Knowledge Test Guide.

   i. "Parts 61/141 Frequently Asked Questions" are located on the Federal Aviation Administration (FAA) Flight Standards Regulatory Support Division (AFS-600) home page on the Internet at: www.mmac.jccbi.gov/afs/afs600/pefaq.html#faq.

4. **PILOT TRAINING AND TESTING.** Part 61 contains aeronautical experience, certification requirements, responsibilities, privileges, and limitations for each grade of certificate: student pilot, pilot, flight instructor, and ground instructor. Under the "total training concept," the areas of operation specified for each grade of certificate by part 61 encompasses the areas of operation and tasks contained in the practical test standards (PTS). Instructors are responsible for training applicants "…to acceptable standards in all subject matter areas, procedures, and maneuvers included in the **tasks** within the appropriate practical test standard." For example:

   a. An applicant for a Private Pilot Certificate must have logged the flight time required by part 61 and have had his/her logbook endorsed by their certificated flight instructor (CFI) who determines that the applicant is proficient in the appropriate areas of operation listed in 14 CFR part 61, §61.107(b).

   b. The flight maneuvers associated with each of the areas of operation listed in §61.107 are found under similar titles in FAA-S-8081-14 (Airplane) — Private Pilot Practical Test Standards and FAA-S-8081-15 (Rotorcraft) — Private Pilot Practical Test Standards. The standards for successful completion of each maneuver and procedure are noted in the elements of each task in the PTS. Each of the maneuvers and procedures listed in FAA-S-8081-14 (Airplane) are discussed and explained in FAA-H-8083-3, Airplane Flying Handbook.

5. **KNOWLEDGE TESTS.** The knowledge tests for the recreational pilot certificate, private pilot certificate, commercial pilot certificate, airline transport pilot (ATP) certificate, flight instructor certificate, ground instructor certificate, and ratings cover the subject areas in which aeronautical knowledge is required by part 61.

   a. If required by §61.35(a)(1), an appropriate knowledge test will only be administered to an applicant who presents acceptable evidence of completion of the required training.

   b. Applicants are not required to show such evidence to take the ATP, CFI, certificated ground instructor (CGI), military competency, or foreign pilot instrument knowledge tests unless they are applying to retake a test after failing that test (per §61.49).

   c. The FAA computerized knowledge test program has been implemented to provide expeditious testing services for applicants at locations and times convenient to the public. For location of computerized testing centers, contact the nearest Flight Standards District Office (FSDO) or use the listing of computerized testing centers on the Internet at: http://www.mmac.jccbi.gov/afs/afs600.

6. **COMPLETION OF GROUND TRAINING OR A HOME STUDY CURRICULUM.** Ground training courses to prepare for the aeronautical knowledge test may be offered by pilot schools, colleges, aviation organizations, and individual flight or ground instructors. Home study curriculums are available from representatives of the aviation industry. Home study curriculums individually developed by students should be compiled from material described in the applicable FAA knowledge test guide. Any one of the following methods may be used by the student to show evidence of ground school or home study curriculum completion.

   a. A certificate of graduation from a pilot training course conducted by an FAA-certificated pilot school, appropriate to the certificate or rating sought, or a statement of accomplishment from the school certifying satisfactory completion of the ground school portion of the course;

   b. An endorsement from an appropriately rated FAA-certificated ground or flight instructor who has certified the applicant has satisfactorily completed the ground training required for the certificate or rating sought and is prepared for the test;

c. A certificate of graduation or statement of accomplishment from a ground school course, appropriate to the certificate and rating sought that was conducted by an agency such as a high school, college, adult education program, the Civil Air Patrol, or a Reserve Officer's Training Corps flight training program; or

d. A certificate of graduation from an industry-provided aviation home study course. The certificate must be developed by the aeronautical enterprise providing the study material. The certificate of graduation must correspond to the FAA knowledge test for the certificate or rating sought. The aeronautical enterprise providing the course of study must also supply a comprehensive knowledge test which can be scored as evidence that the student has completed the course of study. The knowledge test must be sent to the course provider for scoring. Upon satisfactory completion of the examination, a graduation certificate signed by an authorized CFI of the course provider will be provided to the student.

e. Applicants are encouraged to obtain ground training from the sources described above. An applicant who is unable to provide any of the above documents when applying for a knowledge test may present an individually developed home study course to an appropriately rated flight or ground instructor. The instructor will review the course materials and may question the applicant to determine that the course was completed and that the applicant does possess the knowledge required for the certificate or rating sought. The instructor will then complete an endorsement certifying that the applicant is prepared for the knowledge test. See Appendix 1, endorsement No. 43.

7. **EVIDENCE OF IDENTITY AND AGE.** Applicants for knowledge tests and airman certificates must provide positive proof of identification, address, and age at the time of application.

a. **Identification.** The identification presented must include a photograph of the applicant, the applicant's signature, and the applicant's actual residential address (if different from the mailing address). This information may be presented in more than one form of identification. Acceptable methods of identification include, but are not limited to, drivers' licenses, government identification cards, passports, and military identification cards. Some applicants may not possess the identification documentation described. In the case of an applicant under age 21, the applicant's parent or guardian may accompany the applicant and present information attesting to the applicant's identity. If this is done, a statement to that affect will be submitted for enclosure in the airman's permanent record.

b. Address. A temporary mailing address for delivery of the certificate may be indicated on a separate statement attached to the application. However, the address required for official record purposes as shown on an airman application for a certificate must represent the airman's actual permanent residential street address, including apartment number, etc., when appropriate. An alternate mail delivery service address (commercial mail box provider), flight school, airport office, etc., is not acceptable. A post office box or rural route number is not acceptable as permanent residence on an application unless there are unavoidable circumstances that require such an address. An applicant, residing on a rural route, in a boat or mobile (recreational) vehicle, or in some other manner that requires the use of a post office box or rural route number for an address, must attest to the circumstances by signing a statement on a separate sheet of paper. The information provided must include sufficient details to ensure identification of the geographical location of the airman's residence. If necessary to positively identify the place of residence, the applicant may be required to provide a hand-drawn map that clearly shows the location of the residence. When the residence is a boat or other mobile vehicle, the registration number, tag number, etc., and dock or park location must be provided. When applying for the practical test for an airman certificate, a post office address may be specified for use on the certificate issued. A signed request must be submitted with the application for this purpose. The permanent residence address must be shown in the manner specified above.

c. Age. Applicants applying to take a practical test must show they meet the minimum age requirement for the certificate sought. When applying to take a knowledge test, applicants must show that they will meet the minimum age requirement for the certificate sought within 24 calendar months of the date of application for the knowledge test.

8. PRACTICAL TESTS.

a. A practical test is conducted to evaluate the applicant's knowledge and skill for the pilot certificate and rating sought. During a practical test, the examiner will quiz the applicant orally on knowledge elements and ask the applicant to perform the skill elements of the test. However, oral testing may be used at any time during the practical test. An examiner is responsible for determining that the applicant meets the standards outlined in the objectives of each required task in the appropriate PTS.

b. Elements of a maneuver or procedure on the practical test where the applicants are required to be tested orally or by written questions, the PTS uses the words "...the applicant exhibits knowledge of..." or "...the applicant exhibits instructional knowledge of..." in the area of each task noted as "Objective."

c. Elements of a maneuver or procedure on the practical test where the applicants are required to demonstrate their piloting skills, the PTS uses words like considers situations, maintains, utilizes, initiates, transitions to, arrives at, establishes and maintains, remains aware, avoids situations, selects, properly, makes a, recognizes, stops, completes, etc., and similar such words in the area of each task noted as "Objective."

d. In accordance with §61.45(a), a flight simulator or flight training device may be permitted to be utilized for some increments of the practical test. Authorization and the extent of use of a flight simulator or flight training device during the practical test is addressed in the appropriate appendix section of the applicable PTS for the pilot certificate and rating sought.

9. **PREREQUISITES FOR PRACTICAL TESTS.** Except as provided by §61.39(c), each applicant must have received an endorsement from an authorized instructor who certifies the applicant has received and logged the required flight time/training in preparation for the practical test within 60 days preceding the date of the test and has been found proficient to pass the practical test. See Appendix 1, endorsements 12, 18, 20, 22, 24, 37, and 39.

> **NOTE: The endorsement must also state that the applicant has satisfactory knowledge of the subject areas in which he/she was shown to be deficient by the FAA airman knowledge test report, if required. See Appendix 1, endorsements 11, 17, 19, 21, and 23.**

a. All applicants must have the required endorsements specified in part 61 for the aircraft category, class, or rating of certification sought.

b. All applicants who reapply for a retest for a practical test must present another endorsement from their instructor that states the applicant has been given the necessary training and is prepared for the practical test. See Appendix 1, endorsement 36.

c. A practical test, whether or not satisfactorily completed, "uses up" the instructor's endorsement for that test. An instructor's recommendation on FAA Form 8710-1, Airman Certificate and/or Rating Application, is required for each retest conducted for a certificate or rating.

    d. Except for a practical test for a glider category rating or a balloon class rating, applicants must hold at least a current third-class medical certificate.

    e. The ability to read, speak, write, and understand the English language is an eligibility requirement that applies to all pilot certificates and ratings. If the applicant cannot meet this requirement except when such inability to read, speak, write, and understand the English language is due to medical reasons [per §61.13(b) and as allowed by other sections], no certificate or rating will be issued.

10. **STUDENT PILOT CERTIFICATION.** Specific knowledge, flight proficiency, flight experience, and endorsement requirements for the student pilot certificate are located in part 61, subpart C. See Appendix 1, endorsements 1 through 10. A student pilot certificate can be issued by a designated aviation medical examiner as part of a medical certificate. However, an aviation safety inspector (ASI) and designated pilot examiner (DPE) can also issue student pilot certificates. Whenever a student pilot certificate is issued by an ASI or DPE, the applicant must hold a current medical certificate for performing solo privileges in an airplane, rotorcraft, powered-lift, or airship. Glider pilots and balloon pilots are not required to hold a medical certificate. Additional information on the eligibility requirements for student pilots can be found in §61.83 and the general limitations for student pilots can be found in §61.89.

11. **PRESOLO REQUIREMENTS.**

    NOTE: Several questions have been asked about §61.87(l)(1) to clarify that the solo endorsement on the student pilot certificate is a "one-time" endorsement. However, the "90-day" solo endorsement that goes in the student pilot's logbook is required every 90 days for the student to be afforded continuing solo privileges [per §61.87(l)(2)].

    NOTE: Several questions have been asked for clarification on the status of solo endorsements when a person's student pilot certificate has expired. Although §61.19(b) establishes, in pertinent part, that a student pilot certificate expires 24 calendar months from the month it was issued, the endorsements on that student pilot certificate are a matter of record forever. Granted these endorsements are required to be updated from "time-to-time" in the student pilot's logbook to retain solo privileges, but the endorsements on the student pilot certificate are a matter of record indefinitely.

The following presolo requirements must be met:

a. Before being authorized to conduct a solo flight, a student pilot must have received and logged the flight training required by §61.87(c) and (d) through (k), as appropriate. Satisfactory aeronautical knowledge and an acceptable performance level must have been demonstrated to an authorized instructor [per §61.87(b)]. Advisory Circular 61-101 provides information on the required content of the presolo aeronautical knowledge test. See Appendix 1, endorsement 1.

b. Prior to solo flight, a student pilot is required to have his/her student pilot certificate and logbook endorsed for the specific make and model aircraft to be flown. Thereafter, the student pilot's logbook must be endorsed every 90 days to retain solo flight privileges. These endorsements must be given by an authorized flight instructor who has flown with the student [per §61.87(l)]. See Appendix 1, endorsements 2 and 4.

12. ADDITIONAL SOLO PRIVILEGES.

NOTE: For the purpose of ensuring clarification, it has been noted that the student pilot certificate only provides for listing the aircraft's category for the solo cross-country privileges endorsement. Per §61.93(c)(1), the solo cross-country endorsement on the student pilot certificate must be "...for the specific category of aircraft to be flown." However, per §61.93(c)(2)(i), the solo cross-country endorsement in the student pilot's logbook must be "...for the specific make and model of aircraft to be flown."

The following additional student solo privileges may be authorized:

a. A student pilot may operate an aircraft in solo flight at night provided that student has received the required flight training at night and the appropriate endorsements as required by §61.87(m). See Appendix 1, endorsement 3.

b. A student pilot may operate an aircraft on a solo cross-country flight provided that student has received the training required by §61.93(e) through (k) (as appropriate) and has demonstrated acceptable skills, abilities, and competency to his/her instructor who then would endorse the person's student pilot certificate and logbook. Additionally, before each solo cross-country flight, the student's logbook must be endorsed by an instructor [emphasis added "by an instructor" as opposed to the student's instructor, meaning the endorsement does not necessarily need to be the instructor who normally provides training to the student]. However, the instructor who makes the endorsement to authorize this solo cross-country flight will personally review the student's preflight planning and preparation and attest to the correctness and preparedness of the student's cross-country planning under the known circumstances. The instructor may add limitations to the endorsement to ensure an accurate written understanding between the student and the instructor to better ensure the safety of the flight.

c. In the interest of emphasizing the requirements of §61.93(a)(1), when an instructor permits his/her student to make a solo cross-country flight, any solo flight greater than 25 nautical miles (NM) from the airport from where the flight originated, or make any solo flight and landing at any location other than the airport of origination, that student must have received the solo cross-country training and endorsements requirements of §61.93. Additionally, unless there is an emergency, no student may make a solo flight landing at any point other than the airport where the student pilot normally receives training to another location unless that student has received the required solo cross-country training and endorsements of §61.93.

d. A flight instructor may authorize a student to practice solo takeoffs and landings at an airport within 25 NM from the airport at which the student pilot is normally receiving training after meeting the requirements of §61.93(b)(1). See Appendix 1, endorsement 5.

e. A student pilot may be authorized to make repeated, specific solo cross-country flights that are not greater than 50 NM from the point of departure if the student meets the requirements of §61.93(b)(2). The authorized instructor should specify in the student's logbook endorsement, the conditions under which the flights may be made. See Appendix 1, endorsement 8.

f. A student pilot may operate an aircraft in solo flight in Class B airspace or on a solo flight to, from, or at an airport located in Class B airspace provided that student has received the ground and flight training and instructor endorsements required by §61.95(a) and (b). See Appendix 1, endorsements 9 and 10.

13. **RECREATIONAL PILOT CERTIFICATION.** Specific knowledge, flight proficiency, flight experience, and endorsement requirements for the recreational pilot certificate are located in part 61, subpart D. See Appendix 1, endorsements 11 through 16.

a. Section 61.101 contains all limitations that pertain to the recreational pilot certificate and also outlines procedures for obtaining additional certificates or ratings. The training and experience required in furtherance of a higher level of certificate must be supervised by an appropriately authorized flight instructor, and each flight conducted by the recreational pilot under those provisions must be authorized by the flight instructor's endorsement in the recreational pilot's logbook. See Appendix 1, endorsement 16.

b. Recreational pilots may act as pilot in command (PIC) on a flight that is within 50 NM of the departure airport where training was received after having received the required training and endorsement [per §61.101(b)]. Recreational pilots must have their logbook in their personal possession [per §61.101(b)(4)]. See Appendix 1, endorsement 13.

c. Recreational pilots who want to fly beyond 50 NM from the departure airport where training was received must receive additional ground and flight training on the private pilot cross-country training requirements of part 61, subpart E, and must receive an instructor endorsement [per §61.101(c)]. Recreational pilots must have their logbook in their personal possession [per §61.101(c)(3)]. See Appendix 1, endorsement 14.

14. **PRIVATE PILOT CERTIFICATION.** Specific knowledge, flight proficiency, flight experience, and endorsement requirements for the private pilot certificate are located in part 61, subpart E. See Appendix 1, endorsements 17 and 18.

a. An applicant for a private pilot certificate must possess at least a student pilot certificate or a recreational pilot certificate that is current and valid.

b. The flight instructor is given discretion in developing a flight training program to meet the requirements of part 61. The regulation spells out the specific minimum aeronautical experience requirements that must be met.

c. Applicants must meet night experience requirements regardless of medical qualification considerations. The only exception is in accordance with §61.110.

d. Per §61.109(a)(3) and (b)(3), "...3 hours of flight training in a (single-engine airplane) (multiengine airplane) on the control and maneuvering of an airplane solely by reference to instruments, including..."; and §61.109(e)(3), "...3 hours of flight training in a powered-lift on the control and maneuvering of a powered-lift solely by reference to instruments, including..." do not have to be conducted by a certificated flight instructor — instrument (CFII), but the training must be in an aircraft and not in a flight simulator or a flight training device unless it was performed in accordance with a 14 CFR part 142 approved training program.

e. Private pilots (and applicants for the certificate) must understand and comply with all private pilot privileges and limitations, including compensation or hire and expense sharing, in accordance with §§61.113 through 61.117. See Appendix 1, endorsements 17 and 18.

15. COMMERCIAL PILOT CERTIFICATION. Specific knowledge, flight proficiency, flight experience, and endorsement requirements for the commercial pilot certificate is located in part 61, subpart F. See Appendix 1, endorsements 19 and 20.

    a. An applicant for a commercial pilot certificate must hold a private pilot certificate issued under part 61.

    b. An applicant for a commercial pilot certificate with an airplane or powered-lift category rating must hold or concurrently obtain the appropriate instrument rating.

16. AIRLINE TRANSPORT PILOT CERTIFICATION. Specific knowledge, flight proficiency, flight experience, and endorsement requirements for an ATP certificate is located in part 61, subpart G. See Appendix 1, endorsements 40 and 41.

    a. An applicant for an ATP certificate must possess one of the following:

        (1) A commercial pilot certificate and an instrument — (airplane, helicopter, or powered-lift) rating issued under part 61;

        (2) A foreign ATP or a foreign commercial pilot license and an instrument — (airplane, helicopter, or powered-lift) rating, without limitations, issued by a contracting state to the convention on international civil aviation; or

        (3) Be a military pilot or former military pilot and have met the requirements of §61.73 that qualifies the applicant for a commercial pilot certificate with an instrument — (airplane, helicopter, or powered-lift) rating.

    b. An instructor recommendation is not required, unless the applicant has failed the practical test and this is a retest (per §61.49). Applicants for retest must comply with the appropriate retest requirements of §61.49. In addition, §61.49 requires that the instructor sign the applicant's FAA Form 8710-1 for the retest.

17. FLIGHT INSTRUCTOR CERTIFICATION. Specific knowledge, flight proficiency, flight experience, and endorsement requirements for the flight instructor certificate is located in part 61, subpart H. See Appendix 1, endorsements 23 through 26. To be eligible for a flight instructor certificate an applicant must:

    a. Hold a commercial pilot certificate or an ATP certificate with an aircraft rating appropriate to the flight instructor rating sought and was issued under part 61;

    b. Hold instrument rating/instrument privileges on the pilot certificate, if the applicant is seeking a flight instructor certificate with an airplane or powered-lift category rating and was issued under part 61;

c. Have a logbook endorsement certifying that the applicant has been given the required ground and flight training and has been found competent to pass the practical test. The endorsement must be made by an instructor who meets the requirements of §61.195(h). See Appendix 1, endorsements 24 and 25;

d. Have logged at least 15 hours as PIC in the category and class of aircraft that is appropriate to the flight instructor rating sought; and

e. For applicants applying for a flight instructor certificate with airplane and glider category ratings, the applicant must have received a logbook endorsement that attests to satisfactory demonstration of instructional proficiency of stall awareness, spin entry, spins, and spin recovery procedures in airplanes or gliders, as appropriate. A logbook endorsement that attests to satisfactory demonstration of instructional proficiency of stall awareness, spin entry, spins, and spin recovery procedures is required for the initial flight instructor certificate (for a rating in airplanes or gliders). This means, even if the applicant were to initially seek a flight instructor certificate with an airplane multiengine rating, the applicant would still be required to receive a logbook endorsement that attests to satisfactory demonstration of instructional proficiency of stall awareness, spin entry, spins, and spin recovery procedures. However, the training would be required to be performed in an airplane (most likely a single-engine land airplane) that does not contain any restrictions from spins. See Appendix 1, endorsement 26.

18. GOLD SEAL FLIGHT INSTRUCTOR CERTIFICATES.

The specific requirements for the gold seal flight instructor certificate are contained in FAA Orders 8700.1, General Aviation Operations Inspector's Handbook, and 8710.3C, Pilot Examiner's Handbook. Flight instructor certificates bearing distinctive gold seals are issued to flight instructors who have maintained a high level of flight training activity and who meet special criteria. Once issued, a gold seal flight instructor certificate will be reissued each time the instructor's certificate is renewed. Applicants for gold seal flight instructor certificates must meet the following requirements:

a. The flight instructor must hold a commercial pilot certificate with an instrument rating (glider flight instructors need not hold an instrument rating) or an ATP certificate;

b. The flight instructor must hold a ground instructor certificate with an advanced or instrument ground instructor rating; and

c. The flight instructor must have accomplished the following within the previous 24 months:

(1) Trained and recommended at least 10 applicants for a practical test, at least 8 of whom passed their tests on the first attempt;

(2) Conducted at least 20 practical tests as a designated pilot examiner, or graduation tests as chief instructor of a 14 CFR part 141 approved pilot school course; or

(3) A combination of the above requirements. (Two practical tests conducted equal the credit given for one applicant trained and recommended for a practical test.)

19. RENEWAL OF A FLIGHT INSTRUCTOR CERTIFICATE.
The renewal requirements for a flight instructor certificate is located in §61.197. Renewal of a flight instructor certificate may be accomplished at anytime.

a. A flight instructor certificate shall be renewed in accordance with §61.197(a) by accomplishing one of the following methods:

(1) Passing a practical test, that was administered by an examiner, for one of the ratings listed on the person's current flight instructor certificate or passing a practical test for an additional flight instructor rating;

(2) Presenting to an ASI, appropriate records that verify the requirements of §61.197(a)(2)(i) or (ii) were accomplished; or

(3) Presenting to an ASI, evidence of having graduated from an FAA-approved Flight Instructor Refresher Clinic (FIRC) within the preceding 3 calendar months. The instructor's FIRC graduation certificate should be presented to an ASI at the time of application for renewal.

b. In order to comply with §61.197(a)(2)(ii), the FAA offers the following examples of "a position involving the regular evaluation of pilots":

(1) Persons who regularly give aircraft checkouts at a fixed-base operator and the inspector is acquainted with the applicant's duties, responsibilities, and quality of instruction.

(2) 14 CFR parts 121 or 135 airline captains who are "…in a position involving the regular evaluation of pilots…" and have satisfactory knowledge of part 61 pilot training, certification, and standards. The authorized FAA Flight Standards inspector must be acquainted with the duties and responsibilities of those pilot positions.

(3) Company check pilots for 14 CFR part 133 operations who are "…in a position involving the regular evaluation of pilots…" and have satisfactory knowledge of part 61 pilot training, certification, and standards. The authorized FAA Flight Standards inspector must be acquainted with the duties and responsibilities of those pilot positions.

    c. Provided one of the renewal actions of §61.197 takes place within 3 calendar months prior to the expiration month of a current flight instructor certificate, the renewed flight instructor certificate will be valid for an additional 24 calendar months beyond the expiration date shown on the certificate. For example:

        (1) If a flight instructor renewal applicant has the month of August showing on his/her current flight instructor certificate as the expiration month, then that applicant must accomplish one of the flight instructor renewal actions in the months of August, July, June, or May to retain the month of August as the expiration month [per §61.197(b)(2)(i)].

        (2) Another example would be a flight instructor renewal applicant has August 31, 1999, showing on his/her current flight instructor certificate. The applicant completes an FIRC on June 1, 1999, and presents his/her current flight instructor certificate and his/her FIRC graduation certificate to a FSDO on June 1, 1999. The flight instructor certificate is renewed with a new expiration date of August 31, 2001 [per §61.197(b)(2)(ii)].

    d. If the renewal action occurs outside the 3 calendar-month period, the renewed flight instructor certificate will be valid for an additional 24 calendar months from the month the renewal requirements of §61.197(a) were accomplished or in the case of completion of an FIRC when the FIRC graduation certificate is presented to a FSDO. However, the FIRC graduation certificate must be presented to a FSDO within 90 days of completion of an FIRC. For example:

        (1) A flight instructor renewal applicant has August 31, 1999, showing on his/her current flight instructor certificate. The applicant completes an FIRC on March 16, 1999, and presents his/her FIRC graduation certificate to a FSDO on March 16, 1999. The flight instructor certificate is renewed with a new expiration date of March 31, 2001 [per §61.197(b)(1)]. However, the applicant could have taken into consideration the 90-day duration period of the FIRC graduation certificate and held onto it. Taking into consideration the 90-day duration period of the FIRC graduation certificate, the applicant could have presented his/her current flight instructor certificate and his/her FIRC graduation certificate to the FSDO on June 13, 1999 (i.e., the 90th day), and the applicant's flight instructor certificate would have been renewed with a new expiration date of June 30, 2001.

        (2) In either of the previous examples, the renewal action of §61.197(a) must be accomplished prior to the expiration date shown on the applicant's flight instructor certificate.

    e. During the renewal process, when an FIRC graduation certificate is used for renewal and the applicant's flight instructor certificate is being processed for renewal, a copy of the FIRC graduation certificate is considered acceptable documentation by the Administrator until the applicant receives his/her permanent flight instructor certificate. Another example of "...or other documentation acceptable to the Administrator..." [per §61.3(d)], is a copy of an applicant's completed and signed FAA Form 8710-1 that was submitted when making application for renewal.

    f. Notwithstanding the above provisions, any applicant may be required to complete part or all of the applicable flight instructor practical test. A practical test may be required if there is reason to believe that it will serve to correct some deficiency in the applicant's instructing abilities or it has been determined to be necessary to introduce new training procedures or certification requirements.

20. EXPIRED FLIGHT INSTRUCTOR CERTIFICATE. The expired flight instructor certificate requirements are located in §61.199. A practical test is required in the exchange of an expired flight instructor certificate. A designated pilot examiner or an ASI shall not reinstate expired flight instructor certificates without a practical test.

    a. The holder of an expired flight instructor certificate [a flight instructor certificate that conforms to §61.5(c)] may exchange that certificate for a new certificate by passing just one practical test [per §61.183(h)] for one of the aircraft ratings held on that person's expired flight instructor certificate. At the discretion of the examiner, one or all of the ratings held on that person's expired flight instructor certificate may be reinstated.

    b. Flight instructor ratings or limited flight instructor ratings on a pilot certificate are no longer valid. To reinstate instructor privileges, all requirements for initial issuance of a flight instructor certificate must be met.

21. INSTRUMENT RATING. Specific knowledge, flight proficiency, flight experience, and endorsement requirements for the instrument rating are located in §61.65. See Appendix 1, endorsements 21 and 22.

22. ADDITIONAL AIRCRAFT RATINGS (OTHER THAN ATP).

    a. Specific knowledge, flight proficiency, flight experience, and endorsement requirements for additional category, class, or type rating (for other than at the ATP level) are located in §61.63 of subpart B, part 61. See Appendix 1, endorsements 37 through 39.

b. Category and Class Ratings. Applicants adding a category and/or class rating to a pilot certificate must have an instructor's recommendations and appropriate endorsements. An applicant need not take an additional knowledge test, provided the person holds an airplane, powered-lift, rotorcraft, or airship rating at or above the pilot certificate level sought. An applicant must pass the required practical test appropriate to the pilot certificate for the aircraft category and, if applicable, class rating sought. Additionally, applicants must comply with the requirements of §61.63, as noted below:

(1) **Category Ratings.** Must receive the training and have the aeronautical experience required by part 61 that applies to the pilot certificate level for the category, and if applicable, class rating sought.

(2) **Class Ratings.** Must be found competent in the knowledge areas and proficient in the areas of operation for the class rating sought.

(3) **Type Ratings.** Must have the required training time, logbook, or training record endorsements, and pass the required practical test in accordance with the ATP/Type Rating PTS.

(a) The applicant must perform the appropriate areas of operation in actual or simulated instrument conditions unless the aircraft is incapable of such operations. If the aircraft for a type rating is incapable of operating under instrument flight rules, the person may obtain a type rating limited to "VFR **only**" [per §61.63(d)(5) or (h)]. However, the applicant must still comply with §61.63(d)(1), "…Must hold or concurrently obtain an instrument rating that is appropriate to the aircraft category, class, or type rating sought."

(b) An applicant for a type rating to be added to his/her pilot certificate must hold, or concurrently obtain, an instrument rating appropriate to the aircraft category, class, or type rating sought. No instructor recommendation or knowledge test is required for just a type rating when the applicant already holds the aircraft category, class, and instrument rating (appropriate to the rating sought) on his/her pilot certificate. As an example: if a person is applying for a CE-500 type rating at the pilot certificate level held and holds an airplane multiengine land and an instrument-airplane rating, then the person is not required to obtain an instructor recommendation or accomplish a knowledge test. However, the person must comply with §61.63(d)(2) and (3).

(c) If an applicant is concurrently seeking an instrument rating, all the training time, knowledge test, endorsements, and instructor recommendation requirements in accordance with §61.65 are required.

23. ADDITIONAL CATEGORY/CLASS RATINGS AT THE ATP LEVEL.

a. Specific knowledge, flight proficiency, fight experience, and endorsement requirements for additional category, class, or type rating (at the ATP level) are located in §61.165. See Appendix 1, endorsements 40 and 41.

b. An instructor recommendation is not required, unless the applicant has failed the practical test and this is a retest (per §61.49). Applicants for retest must comply with the appropriate retest requirements of §61.49. In addition, §61.49 requires that the instructor sign the applicant's FAA Form 8710-1 for the retest.

24. OTHER INSTRUCTOR ENDORSEMENTS.
Specific requirements for knowledge, aeronautical experience and, as appropriate, testing for the complex airplane, high performance airplane, tailwheel airplane, high altitude/pressurized airplane, and type specific training are found in §61.31. See Appendix 1, endorsements 31 through 34.

25. GROUND INSTRUCTOR CERTIFICATION.
Specific knowledge, flight proficiency, flight experience, and endorsement requirements for the ground instructor certificate is located in part 61, subpart I. See Appendix 1, endorsement 27.

26. AUTHORIZED INSTRUCTORS.
Section 61.1 defines an "authorized instructor" as any instructor who holds a valid and current flight or ground instructor certificate with the applicable privileges and limitations appropriate to the type of instruction provided. Section 61.41 authorizes instructors who are not certificated by FAA to provide training. However, only instructors who are certificated by FAA are allowed to provide the required endorsements toward the requirements for a pilot certificate or rating issued under part 61.

27. ADDITIONAL TRAINING FOR GLIDERS.
Specific knowledge, flight proficiency, and endorsement requirements for the ground tow, aerotow, and self-launch procedures for gliders are located in §61.31(j). See Appendix 1, endorsement 42. With the rewrite of part 61, glider ratings are no longer issued with the "ground tow," "aerotow," and "self-launch" limitations.

a. Pilots who desire to use ground-tow launch procedures must have satisfactorily accomplished ground and flight training on ground-tow procedures and operations. The pilot must have received an endorsement from a CFI-Glider instructor who certifies, in the pilot's logbook, that the pilot has been found proficient in ground-tow procedures and operations.

b. Pilots who desire to use aerotow procedures must have satisfactorily accomplished ground and flight training on aerotow procedures and operations. The pilot must have received an endorsement from a CFI-Glider instructor who certifies, in the pilot's logbook, that the pilot has been found proficient in aerotow procedures and operations.

c. Pilots who desire to use self-launch procedures must have satisfactorily accomplished ground and flight training on self-launch procedures and operations. The pilot must have received an endorsement from a CFI-Glider instructor who certifies, in the pilot's logbook, that the pilot has been found proficient in self-launch procedures and operations.

d. As per §61.31(j)(2), the holder of a glider rating issued prior to August 4, 1997, is considered to be in compliance with the training and logbook endorsement requirements of this paragraph for the specific operating privilege for which the holder is already qualified.

28. HOW TO OBTAIN ELECTRONICALLY.

This AC may be accessed through the Internet temporarily:
    www.faa.gov/avr/afshome.htm
On this web page under General Aviation, click on the link, Certification: Pilots and Flight and Ground Instructors. This AC will remain at this site until it is linked to the Flight Standards Advisory Circular Index, located on the same web page.

L. Nicholas Lacey
Director, Flight Standards Service

Appendix 1 Instructor Endorsements

Contents

No.	*Endorsement Description*	*Page*

Appendix 1
Instructor Endorsements

The following examples are recommended sample endorsements for use by authorized instructors when endorsing logbooks for airmen applying for a knowledge or practical test, or when certifying accomplishment of requirements for pilot operating privileges. Each endorsement must be legible and include the instructor's signature, date of signature, certificated flight instructor (CFI) or certificated ground instructor (CGI) certificate number, and certificate expiration date, if applicable. The purpose for this advisory circular is to provide guidance and to encourage standardization amongst instructors.

Student Pilot Endorsements

1. Presolo aeronautical knowledge: §61.87(b)

I certify that (*First name, MI, Last name*) has satisfactorily completed the presolo knowledge exam of §61.87(b) for the (*make and model aircraft*).

S/S   [date]   J.J. Jones   987654321 CFI   Exp. 12-31-00

2. Presolo flight training: §61.87(c)

I certify that (*First name, MI, Last name*) has received the required presolo training in a (*make and model aircraft*). I have determined he/she has demonstrated the proficiency of §61.87(d) and is proficient to make solo flights in (*make and model aircraft*).

S/S   [date]   J.J. Jones   987654321 CFI   Exp. 12-31-00

3. Presolo flight training at night: §61.87(c) and (m)

I certify that (*First name, MI, Last name*) has received the required presolo training in a (*make and model aircraft*). I have determined he/she has demonstrated the proficiency of §61.87(m) and is proficient to make solo flights at night in a (*make and model aircraft*).

S/S   [date]   J.J. Jones   987654321 CFI   Exp. 12-31-00

4. Solo flight (each additional 90-day period): §61.87(n)

I certify that (*First name, MI, Last name*) has received the required training to qualify for solo flying. I have determined he/she meets the applicable requirements of §61.87(n) and is proficient to make solo flights in (*make and model*).

S/S   [date]   J.J. Jones   987654321 CFI   Exp. 12-31-00

5. **Solo takeoffs and landings at another airport within 25 NM: §61.93(b)(1)**

I certify that (*First name, MI, Last name*) has received the required training of §61.93(b)(1). I have determined that he/she is proficient to practice solo takeoffs and landings at (*airport name*). The takeoffs and landings at (*airport name*) are subject to the following conditions: (List any applicable conditions or limitations.)

S/S [date] J.J. Jones 987654321 CFI Exp. 12-31-00

6. **Initial solo cross-country flight: §61.93(c)(1)**

I certify that (*First name, MI, Last name*) has received the required solo cross-country training. I find he/she has met the applicable requirements of §61.93, and is proficient to make solo cross-country flights in a (*make and model aircraft*).

S/S [date] J.J. Jones 987654321 CFI Exp. 12-31-00

7. **Solo cross-country flight: §61.93(c)(2)**

I have reviewed the cross country planning of (*First name, MI, Last name*). I find the planning and preparation to be correct to make the solo flight from (*location*) to (*destination*) via (*route of flight*) with landings at (*name the airports*) in a (*make and model aircraft*) on (*date*). (List any applicable conditions or limitations.)

S/S [date] J.J. Jones 987654321 CFI Exp. 12-31-00

8. **Repeated solo cross-country flights not more than 50nm from the point of departure: §61.93(b)(2)**

I certify that (*First name, MI, Last name*) has received the required training in both directions between and at both (*airport names*). I have determined that he/she is proficient of §61.93(b)(2) to conduct repeated solo cross-country flights over that route, subject to the following conditions: (List any applicable conditions or limitations.)

S/S [date] J.J. Jones 987654321 CFI Exp. 12-31-00

9. **Solo flight in Class B airspace: §61.95(a)**

I certify that (*First name, MI, Last name*) has received the required training of §61.95(a). I have determined he/she is proficient to conduct solo flights in (*name of Class B*) airspace. (List any applicable conditions or limitations.)

S/S [date] J.J. Jones 987654321 CFI Exp. 12-31-00

10. **Solo flight to, from, or at an airport located in Class B airspace: §§61.95(a) and 91.131(b)(1)**

I certify that (*First name, MI, Last name*) has received the required training of §61.95(a)(1). I have determined that he/she is proficient to conduct solo flight operations at (*name of airport*). (List any applicable conditions or limitations.)

S/S   [date]   J.J. Jones   987654321 CFI   Exp. 12-31-00

Recreational Pilot Endorsements

11. **Aeronautical knowledge test: §§61.35(a)(1) and 61.96(b)(3)**

I certify that (*First name, MI, Last name*) has received the required training of §61.97(b). I have determined that he/she is prepared for the (*name the knowledge test*).

S/S   [date]   J.J. Jones   987654321 CFI   Exp. 12-31-00

12. **Flight proficiency/practical test: §§61.96(b)(5), 61.98(a) and (b), and 61.99**

I certify that (*First name, MI, Last name*) has received the required training of §§61.98(b) and 61.99. I have determined that he/she is prepared for the (*name the practical test*).

S/S   [date]   J.J. Jones   987654321 CFI   Exp. 12-31-00

13. **Recreational pilot to operate within 50 NM of the airport where training was received: §61.101(b)**

I certify that (*First name, MI, Last name*) has received the required training of §61.101(b). I have determined he/she is competent to operate at the (*name of airport*).

S/S   [date]   J.J. Jones   987654321 CFI   Exp. 12-31-00

14. **Recreational pilot to act as PIC on a flight that exceeds 50 NM of the departure airport: §61.101(c)**

I certify that (*First name, MI, Last name*) has received the required cross-country training of §61.101(c). I have determined that he/she is proficient in cross-country flying of part 61, subpart E.

S/S   [date]   J.J. Jones   987654321 CFI   Exp. 12-31-00

15. **Recreational pilot with less than 400 flight hours and not logged PIC time within the preceding 180 days: §61.101(f)**

I certify that (*First name, MI, Last name*) has received the required 180-day recurrent training of §61.101(f) in a (*make and model aircraft*). I have determined him/her proficient to act as PIC of that aircraft.

S/S   [date]   J.J. Jones   987654321 CFI   Exp. 12-31-00

16. **Recreational pilot to conduct solo flights for the purpose of obtaining an additional certificate or rating while under the supervision of an authorized flight instructor: §61.101(i)**

I certify that (*First name, MI, Last name*) has received the required training of §61.87 in a (*make and model aircraft*). I have determined he/she is prepared to conduct a solo flight on (*date*) under the following conditions: (List all conditions which require endorsement, e.g., flight which requires communication with ATC, flight in an aircraft for which the pilot does not hold a category/class rating, etc.).

S/S   [date]   J.J. Jones   987654321 CFI   Exp. 12-31-00

Private Pilot Endorsements

17. **Aeronautical knowledge test: §§61.35(a)(1), 61.103(d), and 61.105**

I certify that (*First name, MI, Last name*) has received the required training of §61.105. I have determined he/she is prepared for the (*name the knowledge test*).

S/S   [date]   J.J. Jones   987654321 CFI   Exp. 12-31-00

18. **Flight proficiency/practical test: §§61.103(f), 61.107(b), and 61.109**

I certify that (*First name, MI, Last name*) has received the required training of §§61.107 and 61.109. I have determined he/she is prepared for the (*name the practical test*).

S/S   [date]   J.J. Jones   987654321 CFI   Exp. 12-31-00

Commercial Pilot Endorsements

19. **Aeronautical knowledge test: §§61.35(a)(1) and 61.123(c)**

I certify that (*First name, MI, Last name*) has received the required training of §61.125. I have determined that he/she is prepared for the (*name the knowledge test*).

S/S   [date]   J.J. Jones   987654321 CFI   Exp. 12-31-00

20. Flight proficiency/practical test: §§61.123(e) and 61.127

I certify that (*First name, MI, Last name*) has received the required training of §§61.127 and 61.129. I have determined he/she is prepared for the (*name the practical test*).

S/S   [date]   J.J. Jones   987654321 CFI   Exp. 12-31-00

Instrument Rating Endorsements

21. Aeronautical knowledge test: §§61.35(a)(1) and 61.65(a) and (b)

I certify that (*First name, MI, Last name*) has received the required training of §61.65(b). I have determined that he/she is prepared for the (*name the knowledge test*).

S/S   [date]   J.J. Jones   987654321 CFI   Exp. 12-31-00

22. Flight proficiency/practical test: §61.65(a)(6)

I certify that (*First name, MI, Last name*) has received the required training of §61.65(c) and (d). I have determined he/she is prepared for the Instrument—(*Airplane, Helicopter, or Powered-lift*) practical test.

S/S   [date]   J.J. Jones   987654321 CFI   Exp. 12-31-00

Flight Instructor Endorsements

23. Fundamentals of instructing knowledge test: §§61.183(d) and 61.185(a)(1)

I certify that (*First name, MI, Last name*) has received the required fundamentals of instruction training of §61.185(a)(1).

S/S   [date]   J.J. Jones   987654321 CFI   Exp. 12-31-00

24. Flight instructor ground and flight proficiency/practical test: §§61.183(g) and 61.187(a) and (b)

I certify that (*First name, MI, Last name*) has received the required training of §61.187(b). I have determined he/she is prepared for the CFI — (*aircraft category and class*) practical test.

S/S   [date]   J.J. Jones   987654321 CFI   Exp. 12-31-00

25. Flight instructor certificate with instrument—(category/class) rating/practical test: §§61.183(g) and 61.187(a) and (b)(7)

I certify that (*First name, MI, Last name*) has received the required CFII training of §61.187(b)(7). I have determined he/she is prepared for the CFII—(*airplane, helicopter, or powered-lift*) practical test.

S/S   [date]   J.J. Jones   987654321 CFI   Exp. 12-31-00

26. Spin training: §61.183(i)(1)

I certify that (*First name, MI, Last name*) has received the required training of §61.183(i). I have determined that he/she is competent and proficient in instructional skills for training stall awareness, spin entry, spins, and spin recovery procedures.

S/S   [date]   J.J. Jones   987654321 CFI   Exp. 12-31-00

> NOTE: The above spin training endorsement is required of flight instructor applicants for the airplane and glider ratings only.

Ground Instructor Endorsement

27. Ground instructor who does not meet the recent experience requirements: §61.217(b)

I certify that (*First name, MI, Last name*) has demonstrated satisfactory proficiency on the appropriate ground instructor knowledge and training subjects of §61.213(a)(3) and (a)(4).

S/S   [date]   J.J. Jones   987654321 CFI   Exp. 12-31-00

[*or CGI, as appropriate]

(The expiration date would apply only to a CFI.)

Additional Endorsements

28. Completion of a flight review: §61.56(a) and (c)

I certify that (*First name, MI, Last name*), (*pilot certificate*), (*certificate number*), has satisfactorily completed a flight review of §61.56(a) on (*date*).

S/S   [date]   J.J. Jones   987654321 CFI   Exp. 12-31-00

> NOTE: No logbook entry reflecting unsatisfactory performance on a flight review is required.

29. Completion of a phase of an FAA-sponsored pilot proficiency award program (WINGS): §61.56(e)

I certify that (*First name, MI, Last name*), (*pilot certificate*), (*certificate number*), has satisfactorily completed Phase No. ___ of a WINGS program on (*date*).

S/S   [date]   J.J. Jones   987654321 CFI   Exp. 12-31-00

30. Completion of an instrument proficiency check: §61.57(d)

I certify that (*First name, MI, Last name*), (*pilot certificate*), (*certificate number*), has satisfactorily completed the instrument proficiency check of §61.57(d) in a (*list make and model of aircraft*) on (*date*).

S/S   [date]   J.J. Jones   987654321 CFI   Exp. 12-31-00

> NOTE: No logbook entry reflecting unsatisfactory performance on an instrument proficiency check is required.

31. To act as PIC in a complex airplane: §61.31(e)

I certify that (*First name, MI, Last name*), (*pilot certificate*), (*certificate number*), has received the required training of §61.31(e) in a (*make and model of complex airplane*). I have determined that he/she is proficient in the operation and systems of a complex airplane.

S/S   [date]   J.J. Jones   987654321 CFI   Exp. 12-31-00

32. To act as PIC in a high performance airplane: §61.31(f)

I certify that (*First name, MI, Last name*), (*pilot certificate*), (*certificate number*), has received the required training of §61.31(f) in a (*make and model of high performance airplane*). I have determined that he/she is proficient in the operation and systems of a high performance airplane.

S/S   [date]   J.J. Jones   987654321 CFI   Exp. 12-31-00

33. To act as PIC in a pressurized aircraft capable of high altitude operations: §61.31(g)

I certify that (*First name, MI, Last name*), (*pilot certificate*), (*certificate number*), has received the required training of §61.31(g) in a (*make and model of pressurized aircraft*). I have determined that he/she is proficient in the operation and systems of a pressurized aircraft.

S/S   [date]   J.J. Jones   987654321 CFI   Exp. 12-31-00

34. To act as PIC in a tailwheel airplane: §61.31(i)

I certify that (*First name, MI, Last name*), (*pilot certificate*), (*certificate number*), has received the required training of §61.31(i) in a (*make and model of tailwheel airplane*). I have determined that he/she is proficient in the operation of a tailwheel airplane.

S/S   [date]   J.J. Jones   987654321 CFI   Exp. 12-31-00

35. To act as PIC of an aircraft in solo operations when the pilot who does not hold an appropriate category/class rating: §61.31(d)(3)

I certify that (*First name, MI, Last name*) has received the training as required by §61.31(d)(3) to serve as a PIC in a (*category and class of aircraft*). I have determined that he/she is prepared to serve as PIC in that (*make and model of aircraft*).

S/S   [date]   J.J. Jones   987654321 CFI   Exp. 12-31-00

36. Retesting after failure of a knowledge or practical test: §61.49.

I certify that (*First name, MI, Last name*) has received the additional (*flight and/or ground*) training as required by §61.49. I have determined that he/she is prepared for the (*name the knowledge/practical test*).

S/S   [date]   J.J. Jones   987654321 CFI   Exp. 12-31-00

> **NOTE: In the case of a failed knowledge test, the instructor may complete the endorsement in the space provided at the bottom of the applicant's airman knowledge test report. The instructor must sign the block provided for the instructor's recommendation on the reverse side of FAA Form 8710-1 application for each retake of a practical test.**

37. Additional aircraft category or class rating (other than ATP): §61.63(b) or (c)

I certify that (*First name, MI, Last name*), (*pilot certificate*), (*certificate number*), has received the required training for an additional (*name the aircraft category/class rating*). I have determined that he/she is prepared for the (*name the practical test*) for the addition of a (*name the aircraft category/class rating*).

S/S   [date]   J.J. Jones   987654321 CFI   Exp. 12-31-00

38. Type rating only, already holds the appropriate category or class rating (other than ATP): §61.63(d)(2) and (3)

I certify that (*First name, MI, Last name*) has received the required training of §61.63(d)(2) and (3) for an addition of a (*name the type rating*).

S/S   [date]   J.J. Jones   987654321 CFI   Exp. 12-31-00

39. Type rating concurrently with an additional category or class rating (other than ATP): §61.63(d)(2) and (3)

I certify that (*First name, MI, Last name*) has received the required training of §61.63(d)(2) and (3) for an addition of a (*name the category/class/type rating*). I have determined that he/she is prepared for the (*name the practical test*) for the addition of a (*name the aircraft category/class/type rating*).

S/S   [date]   J.J. Jones   987654321 CFI   Exp. 12-31-00

40. Type rating only, already holds the appropriate category or class rating (at the ATP level): §61.157(b)(1)

I certify that (*First name, MI, Last name*) has received the required training of §61.157(b)(1) for an addition of a (*name the type rating*).

S/S   [date]   J.J. Jones   987654321 CFI   Exp. 12-31-00

41. Type rating concurrently with an additional category or class rating (at the ATP level): §61.157(b)(1)

I certify that (*First name, MI, Last name*) has received the required training of §61.157(b)(1) for an addition of a (*name the category/ class/type rating*).

S/S   [date]   J.J. Jones   987654321 CFI   Exp. 12-31-00

42. Launch procedures for operating a glider: §61.31(j)

I certify that (*First name, MI, Last name*), (*pilot certificate*), (*certificate number*), has received the required training in a (*list the glider make and model*) for (*list the launch procedure*). I have determined that he/she is proficient in (list the launch procedure).

S/S   [date]   J.J. Jones   987654321 CFI   Exp. 12-31-00

43. Review of a home study curriculum: §61.35(a)(1)

I certify I have reviewed the home study curriculum of (*First name, MI, Last name*). I have determined he/she is prepared for the (*name the knowledge test*).

S/S   [date]   J.J. Jones   987654321 CFI   Exp. 12-31-00

Notes

Notes